I0070918

FM 3-34.400 (FM 5-104)

GENERAL ENGINEERING

December 2008

HEADQUARTERS, DEPARTMENT OF THE ARMY

Field Manual
No. 3-34.400 (5-104)

Headquarters
Department of the Army
Washington, DC, 9 December 2008

General Engineering

Contents

DISTRIBUTION RESTRICTION: Approved for public release; distribution is unlimited.

*This publication supersedes FM 5-104, 12 November 1986.

Figures

Tables

Preface

Field Manual (FM) 3-34.400 is the primary implementing manual for the engineer function that bears its name (the others being combat and geospatial engineering). This FM provides the linkage between the engineering doctrine contained in FM 3-0, FM 3-34, and Joint Publication (JP) 3-34. It specifically draws from the material presented in the Army's keystone engineer manual (FM 3-34) and should always be used with an understanding of its relationship to that manual and its role as the keystone engineer manual. As the implementing manual for the engineer function of general engineering (GE), FM 3-34.400 describes the operational environment (OE) and how to apply and integrate GE principles in support of full spectrum operations and the linkage of GE to assured mobility. This FM focuses on the establishment and maintenance of lines of communications (LOCs) and sustainment operations that support operational requirements throughout the area of operations (AO).

FM 3-34.400 is designed primarily to assist Army engineers at all echelons in planning and coordinating GE operations at the strategic, operational, and tactical levels. It is also a resource applicable to Department of Defense (DOD), joint, and other Army organizations and agencies that have a role in supporting, establishing, and/or maintaining the infrastructure required to conduct and sustain military operations. It is the primary manual to define the engineer function of GE.

FM 3-34.400 is applicable across full spectrum operations. This includes the four types of Army operations (offense, defense, stability, and/or civil support) across the spectrum of conflict (peace, crisis, and war). This FM recognizes the need for joint interdependence and the reality that operations will frequently be performed in a joint, interagency, and multinational environment. This FM describes in detail how to apply the principles of GE when planning and executing GE functions, and is broken down into the following three major parts:

- Part One defines GE in the OE. It provides the staff engineer with the basic concepts and principals necessary to be successful in planning GE missions in support of joint, interagency, and multinational operations.
- Part Two defines the roles and functions associated with gaining and maintaining LOC in support of mobility. It details the responsibilities, planning, and construction/repair actions necessary to assist the force commander in deploying, maneuvering, and redeploying the force.
- Part Three provides information on missions that empower engineers to support sustainment of the force. It includes discussions on procurement of materials, protection support, facilities of various types, base camps, power generation and distribution, well-drilling, and real estate operations.

Although it may be helpful for units conducting construction projects on post, it is not intended to specifically address or focus on the myriad of challenges associated with normal base operations in the continental United States (CONUS) or permanent overseas locations.

The primary audience for FM 3-34.400 is the engineer planner at all echelons. This manual will assist the planner in coordinating, integrating, and synchronizing GE tasks into military operations. GE tasks are part of most military operations. The degree of Army engineer involvement in accomplishing these tasks will vary based on the mission, situation, availability of engineer resources (all Services, host nations [HNs], and contractors), and the commander's intent.

While a dual designated publication, it is intended to inform all Service components of the types of GE tasks, planning considerations, the variety of units available to perform them, and the capabilities of Army engineers to accomplish them. FM 3-34.400 is built directly on the doctrine articulated in—

- FM 3-0.
- FM 3-34.
- JP 3-34.

Planners must recognize that joint and Army transformation is rapidly changing the way we resource and conduct operations, and the application of GE is no exception. The Army has always tailored engineer elements and capabilities to support the force. The provisions of the future engineer force have provided additional modularity into Army engineer organizations to facilitate the commitment of only the required engineer assets into the theater of operations (TO). Enhancing the capabilities of those assets are the reachback capabilities that minimize the footprint of engineers while optimizing the performance of those deployed elements. Planners must apply these improvements and ensure that the GE effort is seamlessly woven into the commander's plan in a proactive fashion and accomplishes the commander's intent.

Terms that have joint or Army definitions are identified in both the glossary and the text. Glossary references: The glossary lists most terms used in FM 3-34.400 that have joint or Army definitions. Terms for which FM 3-34.400 is the proponent FM (the authority) are indicated with an asterisk in the glossary. Text references: Definitions for which FM 3-34.400 is the proponent FM are printed in boldface in the text. These terms and their definitions will be incorporated into the next revision of FM 1-02. For other definitions in the text, the term is italicized, and the number of the proponent FM follows the definition.

Appendix A complies with current Army directives which state that the metric system will be incorporated into all new publications.

This publication applies to the Active Army, the Army National Guard (ARNG)/Army National Guard of the United States (ARNGUS), and the United States Army Reserve (USAR) unless otherwise stated.

The proponent for this publication is the United States Army Training and Doctrine Command (TRADOC). Send comments and recommendations on Department of the Army (DA) Form 2028 (Recommended Changes to Publications and Blank Forms) directly to Commandant, United States Army Engineer School (USAES), ATTN: ATZT-TDD-E, 320 MANSCEN Loop, Suite 220, Fort Leonard Wood, Missouri 64573-8929. Submit an electronic DA Form 2028 or comments and recommendations in the DA Form 2028 format by e-mail to <doctrine.engineer@wood.army.mil>.

Unless this publication states otherwise, masculine nouns and pronouns do not refer exclusively to men.

ACKNOWLEDGMENT

The copyright owners listed below have granted permission to reproduce material from their works. Other sources of quotations, graphics, and material used in examples and vignettes are listed in the Source Notes.

Photograph of a Rubb fabric structure from Rubb Building Systems®. Permission given from the Director of Marketing of Rubb, Inc., 1 Rubb Lane, Sanford, Maine 04073.

Introduction

The three engineer functions are combat (mobility, countermobility, and survivability [M/CM/S]), general, and geospatial engineering. Together, the three functions form the foundation of engineer doctrine, providing the framework for the Engineer Regiment's role in supporting the Army and joint, interagency, and multinational operations. In the past, GE functions have been described almost exclusively as stability operations in a sustainment area. In today's complex OE, it is imperative that GE tasks occur throughout the TO. Engineers must be prepared to perform a full array of GE missions while dealing with a wide range of threats and influences. This FM focuses on engineer command and control (C2), planning, establishment of LOC, and sustainment operations as they pertain to GE. It has applications for engineer leaders and planners at all levels and in all types of engineer units (see FM 3-34 for the unit types and descriptions) that may be conducting GE tasks. While selected GE tasks may be performed by combat engineer units, they are typically performed by GE units (to include the United States Army Corps of Engineers [USACE], other Services, HN, and civilian contactors). Combat engineers are limited from performing GE tasks by their need to focus on combat engineering tasks, lack of organic equipment, and specific training limitations for certain GE tasks.

As an engineer function, FM 3-34.400 is linked to several other manuals. In the joint realm, it is specifically linked to JP 3-34. Within the Army, it is primarily linked to FM 3-34. Additionally, numerous other FMs and technical manuals (TMs) subordinate to the engineer keystone manual provide more depth and technical information concerning each of the discussed chapters (and appendixes) for those requiring more details of the subject areas. As the keystone manual for the engineer function of GE, FM 3-34.400 is the primary source manual for all engineer manuals dealing with the subordinate disciplines, missions, and tasks associated with GE.

GE is the most diverse of the three engineer functions. It occurs throughout the AO, must be planned at all levels of war, is executed during every type of military operation, and is performed by elements of the engineer force from all Services. GE tasks—

- May include, but are not limited to, construction or repair of existing logistics-support facilities, supply and LOC routes (including bridges and roads), airfields, ports, water wells, power generation and distribution, water and fuel pipelines, and base camps/force bed-down. Firefighting and engineer dive operations can be critical enablers to these tasks.
- May be performed by engineer elements of all Services or through the use of other organic means, such as the USACE, Naval Facilities Engineering Command (NAVFAC), or the Air Force Civil Engineering Support Agency (AFCESA).
- May be performed by a combination of joint engineer units, civilian contractors, and HN forces.
- Include the acquisition and disposal of real estate and real property.
- Usually require large amounts of construction materials, which must be planned and provided for in a timely manner.
- May include the production of construction materials.
- Require the integration of environmental considerations. The area of environmental considerations is a subtask under GE in the Army Universal Task List (AUTL).
- Are typically performed by general or construction engineers, but selected GE tasks may also be performed by combat engineers and combat engineer units.

FM 3-34.400 is a significant revision from FM 5-104 in that it reflects the considerable changes that have occurred over the 20 years since that manual was released. While many of the GE tasks have not changed, the OE has shifted. The introduction of field force engineering (FFE) significantly enhanced reachback capabilities and resources, the realities of operations often being joint, interagency, and multinational operations; and the Army's transitional reorganization and restructuring to a modular force has had an impact on doctrine and operations. Changes that directly affect this manual include—

- The advent of the term assured mobility and its relationship to other doctrine. (See FM 3-34.)
- An acknowledgment of the importance of joint interdependence among the Services.
- The introduction of FFE, its relationship to primarily general and geospatial engineering, and the increased integration of the USACE into the integrated support of deployed forces.
- The use of computer-aided planning and management tools.
- The introduction and formalization of a doctrinal process for infrastructure assessment and infrastructure survey as a part of engineer reconnaissance.
- The formalization of a planning tool that supports the engineer staff running estimate known as essential tasks for M/CM/S.
- The likelihood and acknowledgement that most operations conducted will be joint, interagency, and multinational. The primary focus of joint engineer operations is to achieve the commander's intent by coordinating engineer support throughout the joint AO. All branches of Service possess the organic capability to conduct GE. When available, units such as naval mobile construction battalions (NMCBs) (Seabees), Air Force Rapid Engineers Deployable Heavy Operations Repair Squadron, Engineers (RED HORSE), and Prime Base Engineer Emergency Force (Prime BEEF) organizations can greatly increase the GE effort.
- The formalization of support requirements to homeland security. See FM 1, FM 3-07, and JP 3-26.
- The frequency of contractors on the battlefield and their support for many of the GE tasks. (See Army Regulation (AR) 715-9 and FM 3-100.21.)
- The resulting changes in the basic design and organizational structures and equipment of engineer organizations to support the Army's ongoing transformation.
- The acknowledged importance and the requirement to integrate environmental considerations into all operations.

Finally, FM 3-34.400 is written with the acknowledgement that the OE is much more variable than what doctrine was previously written against. Engineers must be prepared to go into any OE and perform its full range of GE tasks while dealing with a wide range of threats and other influences. It builds on the collective knowledge and wisdom gained through recent conduct of operations, numerous exercises, and the deliberate processes of informed reasoning throughout the Army. It is rooted in time-tested principles and fundamentals, while accommodating new technologies and diverse threats to national security.

General Engineering in the Operational Environment

Part one of this manual discusses GE in the OE. It provides guidance for engineers at all levels for integrating and synchronizing GE into the joint theater and maneuver commander's strategic, operational, and tactical plans. Chapter 1 discusses the application of GE as one of the three engineer functions. Chapter 2 provides the fundamentals for the OE in which GE will be applied. Chapters 3 and 4 discuss C2 of engineer operations along with GE planning considerations, to provide a framework to achieve synergy on the battlefield. These are the building blocks for applying GE to the specific GE missions discussed in parts two and three.

Chapter 1

General Engineering as an Engineer Function

Although they were the size of David, engineers did the work of Goliath.
Assistant Division Commander, 101st Airborne Division (Air Assault), Operation Iraqi Freedom After-Action Review

The three engineer functions are combat (M/CM/S), general, and geospatial engineering. As one of three engineer functions, planners integrate the full spectrum of GE to support all warfighter functions at the strategic, operational, and tactical levels. GE encompasses those engineer tasks that establish and maintain the infrastructure required to conduct and sustain military operations. The nature of these tasks requires planners to integrate environmental considerations into the process. Such tasks are conducted in a joint, interagency, and multinational environment and are integrated into the force commander's plan. This force may be led by any one of the Services and GE support may come from any or all Service engineers, contractors, HN capabilities, or the engineers of other nations. This engineer function occurs throughout the AO and across the spectrum of conflict. Past conflicts focused GE on sustainment areas. This may no longer be the case, given the realities of noncontiguous operations against both symmetric and asymmetric threats.

FULL SPECTRUM GENERAL ENGINEERING

1-1. The joint definition says that *GE* is those engineering capabilities and activities, other than combat engineering, that modify, maintain, or protect the physical environment. Examples include construction, repair, maintenance, and operation of infrastructure, facilities, LOCs and bases; terrain modification and repair; and selected explosive hazards (EH) activities. (JP 3-34) This manual serves as the primary reference for planning and executing GE as an engineer function at the Army level. It is directly linked to FM 3-0 and FM 3-34.

1-2. GE is the most diverse of the three engineer functions and is typically the largest percentage of all engineer support provided to an operation. Besides occurring throughout the AO, at all levels of war, and being executed during every type of military operation, it may employ all 23 military occupational specialties (MOSs) within the Engineer Regiment. GE tasks—

- May include construction or repair of existing logistics-support facilities, supply and LOC routes (including bridges and roads), airfields, ports, water wells, power generation and distribution, water and fuel pipelines, and base camps and force bed-down. Firefighting and engineer diving operations are two aspects that may be critical enablers to these tasks.
- May be performed by modified table of organization and equipment (MTOE) units or through the USACE.
- May also be performed by a combination of joint engineer units, civilian contractors, and HN forces, or multinational engineer capabilities.
- Incorporate FFE to leverage all capabilities throughout the Engineer Regiment. This includes the linkages that facilitate engineer reachback (see appendix B).
- May require various types of reconnaissance and assessments to be performed before, or early on in, a particular mission (see FM 3-34.170).
- Include disaster preparedness planning, response, and support to consequence management (CM).
- Include the acquisition and disposal of real estate and real property.
- Include those engineer protection planning and construction tasks that are not considered survivability tasks under combat engineering.
- May include camouflage, concealment, and deception (CCD) tasks (see FM 20-3).
- May include the performance of environmental support engineering missions.
- May include base or area denial missions.
- Usually require large amounts of construction materials, which must be planned and provided for in a timely manner.
- May include the production of construction materials.
- Require the integration of environmental considerations.

1-3. The Chairman of the Joint Chiefs of Staff Manual (CJCSM) 3500.04D contains a hierarchical listing of tasks that are performed by a joint military force. It provides a common language and reference system for joint commanders, staffs, planners, combat developers, and trainers. As applied to joint training, the Universal Joint Task List (UJTL) is a key element of the requirements based mission for task analysis. It contains strategic national and strategic theater tasks, operational tasks, and tactical tasks (theater Army [TA]). Each task also contains measures of performance and criteria that support its definition. At the tactical level, the UJTL links the operational tasks to tactical tasks by requiring Services to produce Service-specific tactical task lists. For the Army this is codified in FM 7-15. Although an analysis of the UJTL is important, most relevant links for GE tasks (since they are typically considered tactical tasks in this hierarchy) are in the AUTL.

> *Note.* The *UJTL* is a menu of capabilities (mission-derived tasks with associated conditions and standards, such as the tools) that may be selected by a joint force commander (JFC) to accomplish the assigned mission. Once identified as essential to mission accomplishment, the tasks are reflected within the command joint mission essential task list. (JP 3-33)

1-4. FM 7-15 outlines GE tasks that units may use as one of the sources to establish their mission-essential task list (METL). Figure 1-1 highlights those Army tactical tasks that are subordinated to providing GE support. While there may be examples of GE tasks not listed under Army tactical task (Provide General Engineer Support), the vast majority are included in these subtasks.

```
┌─────────────────────────────────────────────────────────────────────────┐
│                    ┌──────────────────────────────┐                       │
│                    │      Army Tactical Task       │                       │
│                    │       Provide GE support      │                       │
│                    └──────────────────────────────┘                       │
│                                   │                                        │
│         ┌─────────────────────────┼─────────────────────────┐             │
│ ┌──────────────────┐   ┌──────────────────┐   ┌──────────────────┐       │
│ │ Army Tactical Task│   │ Army Tactical Task│   │ Army Tactical Task│       │
│ │  Restore damaged  │   │  Construct and    │   │  Provide engineer │       │
│ │      areas        │   │  maintain         │   │  construction     │       │
│ │                   │   │  sustainment LOCs │   │  support          │       │
│ └──────────────────┘   └──────────────────┘   └──────────────────┘       │
│                                                                           │
│      ┌──────────────────┐        ┌──────────────────┐                     │
│      │ Army Tactical Task│        │ Army Tactical Task│                     │
│      │  Supply mobile    │        │  Provide          │                     │
│      │  electric power   │        │  facilities       │                     │
│      │                   │        │  engineering      │                     │
│      │                   │        │  support          │                     │
│      └──────────────────┘        └──────────────────┘                     │
└─────────────────────────────────────────────────────────────────────────┘
```

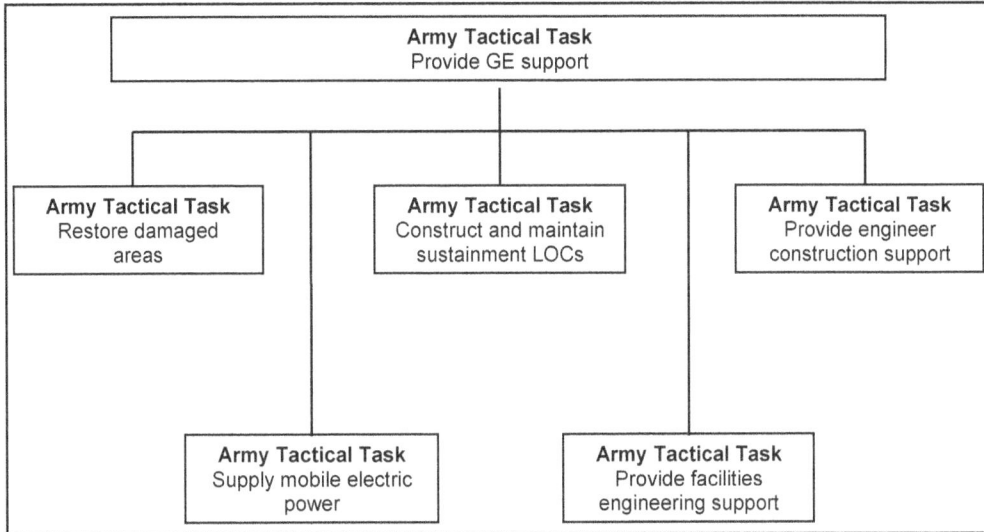

Figure 1-1. GE in the AUTL

1-5.   Engineers conduct GE tasks within the full spectrum of GE operations described in FM 3-0. Within the AO, commanders delineate tasks into two all-encompassing categories of operations (decisive and shaping), thus providing a common focus for all actions.

-   Decisive operations are those that directly accomplish the task assigned by the higher commander and conclusively determine the outcome of major operations, battles, or engagements.

-   Shaping operations create and preserve the conditions for success of decisive operations. Shaping operations enable decisive operations by providing sustainment, sustainment area and base security, movement control, terrain management, and infrastructure development.

1-6.   Commanders organize forces according to purpose by determining whether each unit's operation will be decisive or shaping. GE is usually focused as a shaping operation; however, the commander's intent may dictate that it is at the heart of the decisive operation, particularly in stability or civil support operations. An example of this is provided in the perspective on page 1-4. In the perspective an engineer-led operation was a decisive (or shaping) operation in Afghanistan, with operational, and perhaps even strategic, implications. See the following section for more discussion of GE in full spectrum operations. During execution, the commander combines and directs decisive and shaping operations while preserving opportunities. Ideally, decisive operations occur approximately as planned, while shaping operations create and preserve opportunities and freedom of action to maintain momentum and exploit success.

PERSPECTIVE

During Operation Enduring Freedom V and VI, the rugged, mountainous terrain of Afghanistan's Hindu Kush range became home to Army engineers. The mission was to construct a new, two-lane, 123-kilometer road—a highway from Kandahar City (roughly 25 kilometers northwest of Kandahar Airfield) to Tarin Kowt. Working seven days a week for months in the extreme climate and terrain of Afghanistan, the engineers completed the project ahead of schedule. While this mission was led by engineers from the 528th and 864th Engineer Battalions, it involved an unprecedented level of teamwork between the Army, U.S. government, Afghan National Army, the Afghan government, and various international civilian organizations. The completion of the road marked the end of geographical isolation for hundreds of thousands of Afghan people and assisted the country in its transition toward democracy. This action was more than just the building of a road.

Compilation of Articles

1-7. Commanders visualize their concept of operations and describe their intent. The circumstances may lead the commander to describe their AO in spatial terms of unassigned area, close combat, and sustainment area. These terms may be useful when operations are generally contiguous and against a clearly defined symmetric enemy force. The OE will seldom allow the commander the luxury of describing his AO in such terms. The OE will likely consist of noncontiguous operations against an asymmetric adversary. Figure 1-2 graphically describes possible means by which the commander may visualize his AO.

1-8. The combination of contiguous and noncontiguous operations that the commander uses will have a major impact on the planning and execution of GE tasks. In a contiguous AO, GE tasks are typically performed to the rear of division boundaries by engineer units assigned to higher echelon headquarters. As the AO becomes less contiguous, GE tasks are required in forward areas in proximity to combat units. Since GE assets are not organic to the brigade combat team (BCT), the BCT is normally augmented with the necessary engineer assets to perform GE tasks within the BCT AO. The types of GE assets that will augment the BCT are dependent on the types of missions to be accomplished and the availability of engineers. Selected GE tasks may be performed by combat engineers. The impacts of the noncontiguous battlefield on GE tasks are numerous. They include—

- **The need for increased work site security.** Because units will perform GE near forward elements, contact with the enemy is much more likely. Units conducting GE tasks must be proficient in combat operations to provide for their own defense against such threats. Commanders directing the performance of GE missions must treat these missions as they would any combat operation and ensure the protection of their personnel. General engineers focused on combat operations cannot be focused on performing their GE missions and tasks. It is in the interest of the maneuver commander to keep general engineers out of close combat operations and focused on their GE missions and tasks.

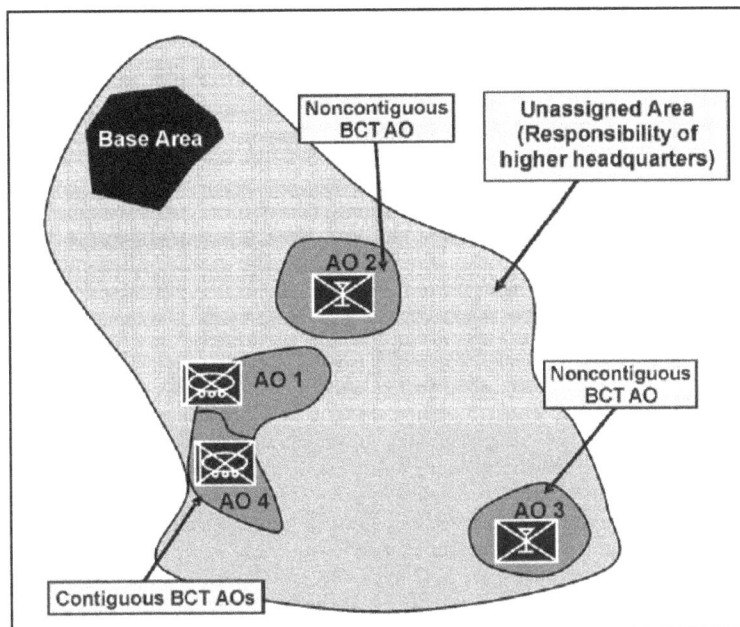

Figure 1-2. Contiguous, noncontiguous, and unassigned areas

- **The need to provide for general and local security.** During contiguous operations, it is often assumed that units receive general security from forward combat units and that only local security is at issue. On the noncontiguous battlefield, units must assume that they will face the same level of threat as maneuver units operating in the AO.

- **The increased number and length of LOCs and main supply routes (MSRs).** With construction and maintenance of these assets critical to sustainment operations, the noncontiguous battlefield greatly expands the GE effort required. Engineer planners can expect smaller-sized units to be spread over greater distances than during contiguous operations. Security of personnel along those routes is an increased concern, and focused convoy security measures will need to be implemented.

- **The need to increase facility construction effort.** Because units will operate with more autonomy within their own AO, they will each require facilities for deployment, supply, maintenance, and other sustainment activities.

- **The increased possibility that combat engineer units may conduct additional GE tasks.** Maneuver commanders at the BCT and higher levels must be able to task their organic combat engineer elements to conduct selected GE tasks. Some tasks can be performed without augmentation. However, a conscious trade-off of potential combat engineering tasks that they could be performing must be made in order to have them perform these tasks. Selected additional GE tasks may be performed when combat engineer units are provided with additional specialized equipment and expertise. Combat engineers will never be able to perform all GE tasks.

- **The likelihood that GE assets will often be task-organized to a much lower level.** Because of the distances involved in a noncontiguous AO, engineer commanders may not be able to effectively C2 the GE effort in a manner that is as responsive to the needs of the maneuver commander without decentralization of authority. These assets may need to be placed in direct support (DS) or attached to BCTs to provide timely and responsive GE support.

EMPLOYMENT CONSIDERATIONS FOR GENERAL ENGINEERING

1-9. Key considerations for the application of GE in an AO include speed, economy, flexibility, decentralization of authority, and establishment of priorities. Effective proactive planning and engineer initiative combine to accomplish the challenges inherent in each of these considerations.

SPEED

1-10. Speed is fundamental to all activities in an AO. Given the tendency for GE tasks to be resource intensive in terms of time, materials, manpower, and equipment, speed is often critical. Proper planning and prioritization are essential to achieve the proper GE effect. Practices that support speed include—

- **Proper prior planning.** Speed is a relative term if planning before the operation did not set the conditions for real speed in terms of mission accomplishment. Speed requires good, broad and inclusive, proactive, and synchronized planning across all staff sections and engineer capabilities.
- **Existing facilities use.** Engineer units must rapidly provide facilities that enable forces to deliver maximum combat power forward. The use of existing facilities may contribute to the essential element of speed by eliminating unnecessary construction support. Use of existing ports, pipelines, warehouses, airfields, and roads during operations is critical. Commanders and staffs must be capable of planning and conducting real estate and real property acquisition to facilitate this effort. Often, the JFC must negotiate with the host government for host nation support (HNS) to use existing facilities. In mature theaters, such as the Republic of Korea, status-of-forces agreements (SOFAs) may dictate procedures for use of existing facilities.
- **Standardization.** Standard materials and plans save time and construction effort. They permit production line methods, including prefabrication of structural members. Standardized assembly and erecting procedures increase the efficiency of work crews by reducing the number of methods and techniques. This supports simplicity. Standardization between all Service engineers is essential for success.
- **Simplification.** Simplicity of design and construction is critical because manpower, materials and time are in short supply. Simple methods and materials allow scarce resources to complete installation in minimum time. They may also allow for the use of HN labor to support the construction.
- **Bare-bones construction use.** Military construction in an AO is characterized by concern for only the minimum necessities by the temporary nature, when possible. The decision on standards to be applied for construction must be decided by the theater commander early in the planning process.
- **Construct in phases.** Phased construction provides for the rapid completion of critical parts of buildings or installations and their use of these parts for their intended purpose before the entire project is complete. The use of the Gantt chart to plan and track progress is a useful tool.

ECONOMY

1-11. GE in an AO demands efficient use of personnel, equipment, and materials. Proper proactive planning is the first step in any discussion of the application of economy.

- **Conserve manpower.** Construction tasks are time-consuming and engineer commanders will often find a shortage of engineers and construction workers. Conservation of labor is important and every engineer must function at the peak of efficiency for long hours to accomplish the GE mission. Careful planning and coordination of personnel is necessary. Missions must be well organized and supervised and personnel carefully allocated for the task. Selected GE tasks may also be performed by combat engineers, but this will involve a conscious decision by the commander to trade off one set of engineer capabilities to further GE tasks.
- **Conserve equipment.** Military construction equipment will likely be in short supply at the beginning of a contingency operation. Operational capability of equipment may be impaired by shortages in repair parts and maintenance personnel. Engineer commanders should consider

contracting for local equipment and repair parts to alleviate these shortages. Preventive maintenance of equipment is essential to ensure that it is available for long-term use. Commanders must ensure that time is allocated for scheduled services to optimize equipment capabilities.

- **Conserve materials.** The critical aspect of completing a GE task is often the availability of appropriate materials. Although planners should make maximum use of local resources in the area of responsibility (AOR) to maximum extent possible, they may not be available, or in short supply. If this is the case, planners must anticipate the need to ship materials from outside the AO, taking into consideration the long transit times that may be required. Conservation of materials while executing GE tasks should always be a critical consideration.

- **Apply environmental considerations early in the process.** While some situations require putting aside risk associated with environmental considerations, it is generally true that the earlier the risk is mitigated, the easier and less complex mitigation procedures will need to be employed (and at the least cost). As the staff proponent for environmental issues, engineers must be able to analyze environmental considerations and recommend appropriate courses of action (COAs) to the commander. If an environmental baseline survey (EBS) and an environmental health site assessment (EHSA) will be required, see that they are performed early in the process.

Note. An *EBS* is a coordinated boundary or phase line used to compartmentalize an AO to indicate where specific engineer units have primary responsibility for the engineer effort. It may be used at division level to discriminate between an AO supported by division engineer assets and an AO supported by DS or general support (GS) corps engineer units. (FM 5-100)

- **Identify funding.** The types of funding and considerations involved are identified in FM 3-34 and JP 3-34. Identifying appropriate funding and proper use of these funds is an important application of economy.

FLEXIBILITY

1-12. The rapidly changing situation during operations requires that GE tasks in all stages be adaptable to new conditions. To meet this requirement, use standard plans that allow for adjustment, expansion, and contraction whenever possible. For example, a standard building plan may be easily adapted for use as an office, barracks, hospital ward, or dining facility. Forward airfields should be designed and located so that they can be expanded into more robust facilities capable of handling larger aircraft and a larger maximum (aircraft) on ground (MOG) capacity. Standardization enhances flexibility.

1-13. Flexibility also implies versatility between Service engineers and within engineer organizations to accomplish GE tasks. This may include providing selected technical expertise and equipment to a variety of engineer organizations to perform GE missions that they are not specifically designed for or organized to perform. Engineer units must display a multifunctional ability to perform engineer tasks outside of their METL. An example of this might be the use of engineers that typically perform combat engineering tasks to perform selected GE tasks. A decision like this would require a risk analysis and the approval of the higher echelon commander to ensure that those engineers are not taken away from other more critical missions in their role of providing support to movement and maneuver for BCTs and other combat forces.

1-14. The basic deployability of engineer organizations and the modularity built into their designs are enablers of flexibility. Engineers must be ready to send only those assets specifically required to perform a mission and establish functional high-performing teams from a variety of Army engineer units and capabilities and from multiple Service engineer organizations. Integration of commercial engineer equipment and flexibility of engineer C2 must support the consideration of decentralization of authority.

DECENTRALIZATION OF AUTHORITY

1-15. The wide dispersion of forces in the AO requires that engineer authority be decentralized as much as possible. The engineer commander or engineer coordinator (ENCOORD) in charge of operations at a particular location must have authority consistent with his responsibilities. As noted before, this is essential on the noncontiguous battlefield.

1-16. Decentralization of authority requires effective C2 and flexibility of its application to integrate the variety of engineer capabilities that may be employed to accomplish selected GE tasks or missions. Service engineers must be capable of nearly seamless integration between units and capabilities to meet the needs of the joint or component commander.

ESTABLISHMENT OF PRIORITIES

1-17. It is essential to establish priorities to determine how much GE effort must be devoted to a single task. While detailed priority systems are normally the concern of lower echelon commands, all levels beginning with the JFC and Army service component commander (ASCC) must issue directives establishing broad priority systems to serve as a guide for detailed systems. Resources must initially be assigned only to the highest priority tasks. Low priority tasks may be left undone while recognizing and assessing the risk of doing so. At theater level, planners can assume general priorities for initial phases of an operation and refine these priorities as the planning effort matures.

ASSURED MOBILITY INTEGRATION

1-18. While focused primarily on the warfighting functions of movement and maneuver, intelligence, and protection, assured mobility has linkages to each of the warfighting functions and both enables and is enabled by those functions. While the engineer has a primary staff role in assured mobility, other staff members support its integration and have critical roles to play. The engineer plays an integrating role in assured mobility that is similar to the role played by the intelligence officer in the intelligence preparation of the battlefield (IPB) integrating process. Ultimately, assured mobility is the commander's responsibility. Other staff members also integrate M/CM/S tasks as a part of assured mobility. Examples include the regulation of traffic in the maneuver space, the handling of displaced persons, and other M/CM/S tasks to support the maneuver plan. Assured mobility is the integrating planning process within which consideration of engineer; chemical, biological, radiological, and nuclear (CBRN); and other reconnaissance capabilities will occur.

1-19. The framework of assured mobility follows a continuous cycle of planning, preparing, and executing decisive and shaping operations. Achieving assured mobility rests on applying six fundamentals that sustain friendly maneuver, preclude the enemy's ability to maneuver, and assist the protection of the force. The fundamentals of assured mobility and some of the specific linkages to GE are to—

- **Predict.** Engineers and other planners must accurately predict potential enemy impediments to joint force mobility by analyzing the enemy's tactics, techniques, procedures, capability, and evolution. Prediction requires a constantly updated understanding of the OE. Applying GE, planners must predict the impact of enemy operations and military operations on the infrastructure required to maintain mobility and momentum. An example would be predicting the damage to a MSR caused by the movement of a large mechanized force over a single route.

- **Detect.** Using intelligence, surveillance, and reconnaissance (ISR) assets, engineers and other planners identify early indicators for the location of natural and man-made obstacles, preparations to create and emplace obstacles, and a potential means for obstacle creation. They identify both actual and potential obstacles and propose solutions and alternate COAs to minimize or eliminate their potential effects. For the GE function, planners must detect impacts to strategic, operational, and tactical mobility that can be affected by engineering solutions.

- **Prevent.** Engineers and other planners apply this fundamental by denying the enemy's ability to influence mobility. This is accomplished by forces acting proactively before the obstacles are emplaced or activated. This may include aggressive action to destroy enemy assets and capabilities before they can be used to create obstacles. Political considerations and rules of engagement (ROE) may hinder the ability to apply the fundamental early in a contingency. Commanders must apply the necessary GE assets in a timely manner to prevent mobility impediments to the force. This may include such actions as construction of a bridge bypass before a bridge becomes unusable.

- **Avoid.** If prevention fails, the commander will maneuver forces to avoid impediments to mobility if this is viable within the scheme of maneuver. GE may become an integral part of the maneuver force's ability to avoid such impediments. Engineers may be required to build roads around natural or man-made obstacles, construct alternate airfields, and other actions that allow maneuver elements to operate effectively.

- **Neutralize.** Engineers and other planners plan to neutralize, reduce, or overcome obstacles and impediments as soon as possible to allow unrestricted movement of forces. The breaching tenents and fundamentals apply to the fundamental of "neutralize. Building a tactical or LOC bridge may be an example of a GE task that neutralizes a river obstacle.

- **Protect.** Engineers and other elements plan and implement survivability and other protection measures that will deny the enemy the ability to inflict damage as joint forces maneuver. This may include countermobility missions to deny the enemy maneuver and provide protection to friendly maneuvering forces. Commanders can direct that GE efforts focus on survivability support such as berms, bunkers, and hardened facilities. This is primarily focused on the hardening aspect of survivability as described in FM 5-103.

Note. Survivability is the concept which includes all aspects of protecting personnel, weapons, and supplies while simultaneously deceiving the enemy. Survivability tactics include building a good defense; employing frequent movement; using concealment, deception, and camouflage; and constructing fighting and protective positions for both individuals and equipment. The Army definition adds, "Encompasses planning and locating position sites, designing adequate overhead cover, analyzing terrain conditions and construction materials, selecting excavation methods, and countering the effects of direct and indirect fire weapons." See FM 5-103.

1-20. Assured mobility provides the broad framework of fundamentals that serve to retain the focus and integrate M/CM/S within the combined arms team. Planners at all levels of the combined arms team rely on this framework to ensure that adequate support is provided to the commander's scheme of maneuver and intent. Within the combined arms team planning staff, it is the assured mobility section at the BCT level (and those same staff members at echelons above the BCT) that provides input for engineer, CBRN, and similar specialized reconnaissance. The ENCOORD plans for the application of and coordinates the integration of engineer reconnaissance across the engineer functions and spans the range from tactical to technical capabilities.

1-21. GE facilitates the force's ability to apply the fundamentals of assured mobility. As engineers plan engineering support as part of an operation, they must integrate each engineer function, including GE, into the operational context necessary to support assured mobility and the maneuver commander.

1-22. Engineer units with primarily a GE mission must be trained and prepared to execute all engineer functions to support the maneuver commander. They must be able to use and integrate geospatial products into their operations and conduct limited combat engineering functions to facilitate their GE mission. They must also be well trained in small-unit tactics, to include convoy security, work site defenses, and limited offensive operations.

FULL SPECTRUM OPERATIONS

1-23. The Army's operation concept is full spectrum operations (see figure 1-3, page 1-10, and FM 3-0). Full spectrum operations are the purposeful, continuous, and simultaneous combinations of offense, defense, and stability or civil support to dominate the military situation at operational and tactical levels. In full spectrum operations, Army forces adapt to the requirements of the OE and conduct operations within it using synchronized action, joint interdependent capabilities, and mission command. They defeat adversaries on land using offensive and defensive operations, and operate with the populace and civil authorities in the AOs using stability or civil support operations.

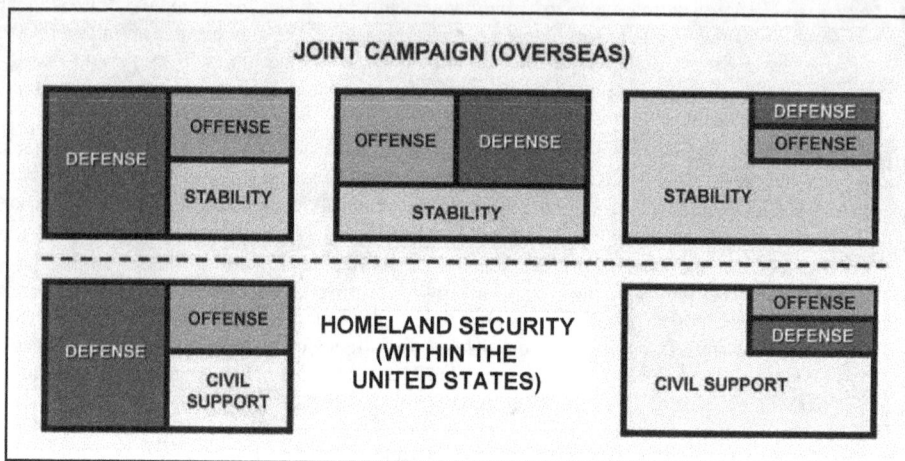

Figure 1-3. Full spectrum operations

1-24. The Engineer Regiment is organized and equipped to respond to the broad range of full spectrum operations. In spite of this, engineers can expect serious challenges in the OE when trying to execute GE tasks. A lack of resources—to include equipment, personnel, and logistics—may severely impede the commander from executing all necessary tasks, and careful prioritization must occur. Even more challenging is that in the OE, units must be able to rapidly transition from the different types of operations. Units supporting an offensive operation must be able to rapidly transition to defense or stability operations. Likewise, a contingency that begins as a stability operation may rapidly move to defense or the offense. This may occur at the strategic, operational, and tactical levels, and engineers apply the GE principle of flexibility to facilitate this transition. This makes executing GE tasks all the more challenging in the OE because of the long duration GE tasks often require.

1-25. To most effectively accomplish the tasks assigned to engineers, it is necessary that commanders carefully consider augmentation requirements. GE units are very capable of accomplishing their assigned tasks; however, they are designed to accomplish specific types of tasks. Therefore, it is imperative that when task-organizing engineers, proper assets be allocated from the engineer force pool.

OFFENSIVE OPERATIONS

1-26. Offensive operations carry the fight to the enemy by closing with and destroying enemy forces, seizing territory and vital resources, and imposing the commander's will on the enemy. They are focused on seizing, retaining, and exploiting the initiative. Assured mobility supports all the forms of offensive maneuver including the envelopment, turning movement, penetration, infiltration, and frontal attack. Executing the forms of maneuver translates into four types of offensive operations at the tactical level: the movement to contact, attack, exploit, and pursuit. See FM 3-90 for an in-depth discussion of these forms and types of offensive maneuver. The primary focus of the GE function is reinforcing combat engineer support to operational and tactical maneuvers and infrastructure support focused on the creation and sustainment of LOCs.

1-27. Although planners must anticipate actual requirements based on a thorough and continuous engineer estimate, offensive operations will likely require the execution of many of the engineering missions below. Several of these are combat engineering missions and tasks, but may be performed by units that most typically perform GE tasks. They include—

- Constructing and repairing roads and combat trails used as supply routes.
- Ensuring theater access through the construction and upgrade of ports, airfields, and reception, staging, onward movement, and integration (RSO&I) facilities.

- Including the repair of paved, asphalt, and concrete runways and airfields as part of forward aviation combat engineering.
- Installing assets that prevent foreign object damage (FOD) to rotary wing aircraft.
- Constructing tactical and LOC bridging.
- Conducting area damage control (ADC) missions that support the mobility of the maneuver force.
- Constructing internment/resettlement (I/R) facilities.

DEFENSIVE OPERATIONS

1-28. Defensive operations defeat enemy attacks, buy time, economize forces, and develop conditions favorable for offensive operations. Defensive operations alone cannot achieve a decisive victory. Their purpose is to create conditions for a counteroffensive that will regain the initiative. There are three types of defensive operations: area defense, mobile defense, and retrograde (see FM 3-90). Although the primary focus for engineers is on combat engineering to enable combined arms obstacle integration and assured mobility to friendly repositioning or counterattacking forces, GE tasks will still play an important role. Examples of expected missions include—

- Constructing hardened facilities that protect the force from enemy artillery and air attack.
- Reinforcing combat engineer efforts in M/CM/S.
- Constructing and repairing of routes that facilitate the repositioning of forces throughout the AO.
- Repairing ports, railroads, pipelines, and other assets critical to sustainment operations.
- Constructing decontamination facilities.

STABILITY OPERATIONS

1-29. Stability operations promote and protect United States (U.S.) national interests by influencing diplomatic, civil, and military environments. They are conducted as a part of overseas operation. Army forces support stability operations by sustaining and exploiting security and control over areas, populations, and resources. They act as part of a joint force with the U.S. country team and the United States Agency for International Development (USAID). GE tasks primarily focus on the reconstruction or establishment of services that support the population in conjunction with civilian agencies in addition to their normal support of U.S. forces. Engineers conducting missions provide resources to assist in disaster or theater response in areas outside U.S. territory. Rapid and effectively emplaced sustainment operations reduce human injuries and fatalities and harden infrastructure. Regional security is supported by a balanced approach that simultaneously enhances regional stability and economic prosperity. FM 3-07 discusses the types of operations that Army forces may conduct during stability operations. Engineers are focused on assisting in stabilizing a region by improving the infrastructure and integrating with and supporting maneuver forces in their missions. The majority of the overall engineer effort during stability operations is likely to be through the GE function. Given the nature of stability operations, the risks associated with environmental hazards may have a greater importance and impact on stability operations than on offensive or defensive operations.

1-30. Stability operations tend to be a long duration compared to the other full spectrum operations. As such, the GE level of effort is very high at the onset and gradually decreases as the theater matures. As the AO matures, GE effort may transfer to civilian contractors, such as those that operate under the logistics civilian augmentation program (Army). Because of the recognition that U.S. forces will likely remain long term, the GE missions tend to focus on the long-term sustainment of the force. Likely missions include—

- Base camp and force bed-down facility construction.
- Survivability and other protection support.
- Robust support area facilities.
- Infrastructure support.
- Power generation and distribution facilities that are reliable.
- LOC construction, maintenance, and repair.

1-31. Stability operations often include humanitarian and civic assistance (HCA). By law (Section 401, Title 10, United States Code [USC], HCA is authorized by the secretary of state and appropriated in the Army budget. In foreign humanitarian assistance operations, Army forces act as part of a joint force with the U.S. country team and the USAID. A reconnaissance (assessment and survey) of the local infrastructure is an important part of stability operations and should complement the humanitarian effort. This reconnaissance is one of the first steps in determining priorities for infrastructure improvement for the local population (see chapter 4 and FM 3-34.170).

CIVIL SUPPORT OPERATIONS

1-32. Civil support operations provide essential services, assets, or specialized resources to assist civil authorities dealing with situations, both natural and man-made, that are beyond their capabilities within the United States and its territories. These operations often involve a variety of actions that directly provide governmental agencies and nongovernmental organizations (NGOs) with support to operations to alleviate hunger, disease, or other consequences of a man-made or natural disaster. A presidential declaration of a national emergency is required in most instances for active duty forces to participate in civil support operations. For additional information see FM 3-07.

1-33. Civil support operations are only performed within the United States and its territories. They, along with offensive and defensive operations, are the three types of operations performed by Army forces in support of homeland security (Figure 1-4). There are very few new or unique GE missions performed in support of homeland security that are not performed during other operations. The difference is the context in which they are performed.

1-34. The primary missions for engineers during civil support operations reside in the areas of protection, CM, and community assistance (civil affairs [CA]) projects. The protection and potential mitigation and recovery of defense critical infrastructure can include both DOD and non-DOD entities. This infrastructure is defined as DOD and non-DOD cyber and physical assets and associated infrastructure essential to project and support military forces worldwide. They could include selected civil and commercial infrastructures that provide the power, communications, transportation, and other utilities that military forces and DOD support organizations rely on to meet operational needs.

HOMELAND SECURITY IMPLICATIONS FOR GENERAL ENGINEERING

1-35. Military forces conduct operations to accomplish missions in both war and peace. Within the United States and its territories, Army forces support homeland security operations. Homeland security operations provide the nation with strategic flexibility by protecting its citizens and infrastructure from conventional and unconventional threats. Homeland security has two major components: homeland defense (primarily offensive and defensive operations) and civil support (primarily domestic emergencies). The discussion below articulates the parameters assigned to the military when performing homeland defense (figure 1-4).

Homeland Security
Homeland security, is defined a concerted national effort to prevent terrorist attacks within the U.S.; reduce America's vulnerability to terrorism, major disasters, and other emergencies; and minimize the damage and recover from attacks, major disasters, and other emergencies that occur. (JP 3-28)

Homeland Defense
Homeland defense is the protection of U.S. sovereignty, territory, domestic population, and critical defense infrastructure against external threats and aggression (this definition was shortened, and the complete definition is printed in the glossary). (JP 3-27)

Civil Support
Civil support is defined as DOD support to U.S. civil authorities for domestic emergencies, and for designated law enforcement and other activities. (JP 3-26)

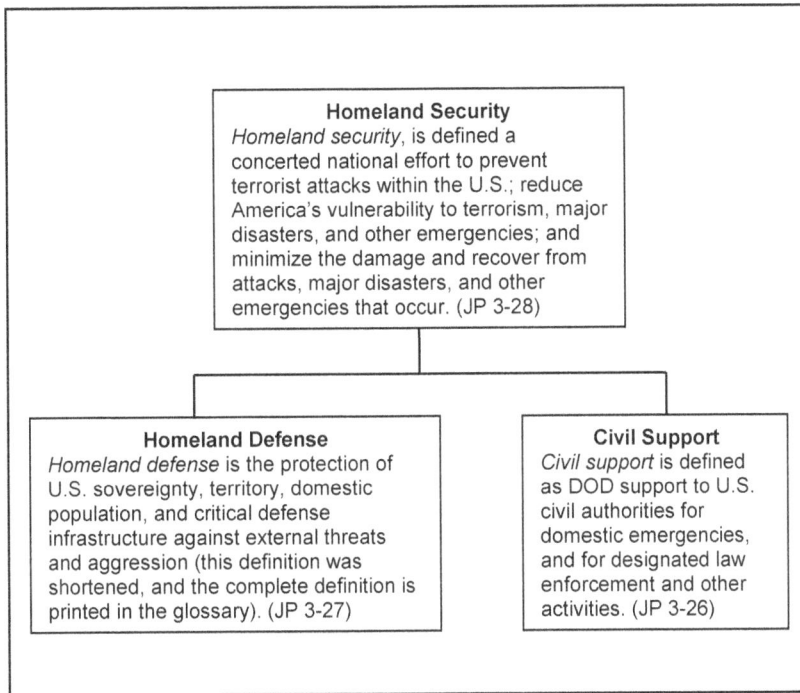

Figure 1-4. Operational descriptions of homeland security and mission areas

1-36. Under extraordinary circumstances and when directed by the proper authority, military forces may conduct offensive and defensive combat operations within the sovereign territory of the United States to prevent, deter, preempt, and defeat an enemy attack.

1-37. Army forces conduct homeland security operations as directed as part of a JFC under the United States Northern Command (USNORTHCOM). JP 3-26 provides the fundamental principles and overarching guidance for the armed forces in homeland defense. It describes the homeland security framework and supporting missions, legal authorities, joint and interagency relationships, and C2. The DOD is the lead federal agency and USACE is the DOD lead agent for the emergency support function of public works and engineering.

1-38. Despite the doctrinal voids, engineers must still be capable of executing GE missions as part of homeland defense. Careful mission analysis is critical to success during these operations. Engineer planners must be aware of the statutes and regulations that restrict military interaction with other government agencies and civilians. Statutory restrictions in the form of the Posse Commitatus and Stafford Acts require commanders to seek legal advice from the Staff Judge Advocate (SJA) early in the planning process.

1-39. Homeland defense may integrate GE to shape the AO. The primary supporting tasks fall into the following three categories:

- **Prevent.** Deny access and interdict territorial intrusion and specified illegal activities. Improve physical security measures by constructing hardened facilities.
- **Protect.** Enhance protection and antiterrorism (AT) by restricting access to installations, facilities, equipment, and material. Protection is a key enabler of prevention.
- **Respond.** Engage populations and mitigate effects. GE assets may be used for debris removal, utilities restoration, facilities repair, temporary shelter construction, and road and airfield repair.

1-40. Engineers plan and execute GE tasks to support critical infrastructure defense and protection missions as they are assigned in operational plans and orders. Security-related operations and protection measures are the responsibility of all elements of the combined arms team.

1-41. The following items should be considered when establishing policy, formulating plans, or analyzing sites in homeland defense:

- Specific areas (facilities) and items to be secured.
- Anticipated threat.
- Degree of protection required.
- Priority for preparation and execution.
- Command channels that will apply for the specific site.
- Assignment of planning and execution responsibility.
- Assistance to be provided or desired for protecting the targets from enemy interference.
- Safety and security measures to be followed.
- Federal, state, and local environmental laws and regulations.
- National policy restrictions.
- Coordination required between military elements and civil authorities.
- Allocations of available and local resources. Base camp planning board.

Chapter 2

Operational Environment

Above all, we must realize that no arsenal, or no weapon in the arsenals of the wicked are so formidable as the will and moral courage of free men and women. It is a weapon adversaries in today's world do not have.

Ronald Reagan

The *OE* is a composite of the conditions, circumstances, and influences that affect the employment of military forces and bear on decisions of the commander (JP 3-0). FM 3-0 outlines the dimensions of the OE as the threat, political, unified action, land combat operation, information, and technology. Commanders tailor forces, employ diverse capabilities, and support different missions to succeed in this complex environment. FM 3-34 further defines the OE in terms of how it affects the employment of the Engineer Regiment's capabilities. The purpose of this chapter is not to repeat the information from FM 3-0 and FM 3-34 but to describe specifically major areas of the OE and unified action that have distinct impacts of GE efforts; warranting further discussion here.

OPERATIONAL ENVIRONMENT

2-1. Although the range of threats and conditions during this period extends from the small, low-technology opponents using adaptive asymmetric methods to larger mechanized forces able to engage deployed U.S. forces in symmetric manners, the OE is constantly evolving and is highly dependent on the region and specific threat.

2-2. Engineers must continuously evaluate the OE to prepare for the requirements of full spectrum operations. An examination of the operational variables that describe the OE is necessary to understand the requirements. The Center for Army Lessons Learned (CALL) is responsible for maintaining the OE handbooks, which outlines current prevailing thought. Engineers should analyze these OE handbooks apply this information to training scenarios.

THREAT IN THE OPERATIONAL ENVIRONMENT

2-3. The threat is any specific foreign nation or organization with intentions and military capabilities that suggest it could be adversarial or challenge the security interests of the United States, its friends, or its allies. Any potential adversary of the United States can be defined as a threat.

2-4. In the OE, the threat is varied and diverse. It may involve participants that are nation states and non-nation states. Nation states may consist of major powers, transition states, rogue nations, failed states, and countries that rapidly change their status. Nation states that are adversarial may form coalitions that impact the analysis of the threat. Non-nation actors include terrorist, drug-trafficking, and criminal organizations.

2-5. Engineers must carefully analyze how the threat tends to view the United States' predominant views today, including that the United States is—
- A major power with overall technological advantages.
- Reliant on air operations and standoff technology and will avoid direct fighting.
- Dependent on information dominance.

- Unwilling to accept heavy losses.
- Sensitive to domestic and world opinion and lacks commitment over time.
- Lacking cultural awareness.
- Predictable in conducting military operations.
- Vulnerable because of our coalitions, force projection, dependence on robust logistics, reliance on contractor support, and dependence on critical resources.

2-6.   Engineers must carefully analyze the threat to adequately predict and plan the GE requirements for an operational mission. Changes in the OE threat may impact how engineers equip and train units designed to perform primarily GE missions. Planners should use their knowledge of the OE in a specific AO to predict the level of GE effort by examining the state of the infrastructure, to include ports, roads, bridges, airfields, and utilities. An examination of the physical environment, to include environmental considerations impacting the mission, will assist in the determination of the level and type of requirements.

2-7.   Recent operations have demonstrated how various threats have influenced the GE effort. In Operation Iraqi Freedom, an enemy that is less capable of militarily confronting U.S. forces resorts to attacking infrastructure to disrupt coalition efforts. This requires engineers to plan, resource, and execute GE tasks across the OE that were not seriously envisioned in past operating environments.

UNIFIED ACTION

2-8.   As part of the OE, unified action describes the wide scope of actions that take place within unified commands, subunified commands, and joint task force (JTF) AORs to achieve national and military objectives. Unified action is the term that describes the integration of joint, Service, interagency, multinational, and nongovernmental efforts to achieve common purpose. Unified action is a recognition that all resources available are brought to bear to achieve the combatant commander's (CCDRs) objectives. The ultimate objective of unified action is unity of effort among many diverse agencies in complex OEs.

2-9.   Army forces do not fight alone; they fight as part of a joint team. A joint force is composed of forces of two or more Services under a single JFC. Effective joint integration does not demand joint commands at all echelons but does require an understanding of joint interdependence at all echelons. Joint interdependence combines Army forces' strengths with those of other Services. The combination of multiple and diverse joint force capabilities creates military effects more potent than those produced by any Service alone.

2-10. Joint interdependence reinforces and complements the effects of Army combined arms operations and makes Army forces many times more effective than they would be otherwise. It provides capabilities not available to any single Service, increasing the force's speed, shock, surprise, depth, simultaneity, and endurance, thereby enabling the Army's operational concept and full spectrum operations as a whole. Joint force capabilities provide additional mobility, intelligence, fires, protection, and logistics throughout the land AO.

2-11. At the strategic level, joint interdependence allows each Service to divest itself of redundant functions that another Service provides better. Doing this reduces unnecessary duplication of capabilities among the Services. It achieves greater efficiency in all areas of expertise. Interdependence allows the Army to focus on developing capabilities that only land forces can provide. Likewise, relying on the Army for land-related capabilities allows the other Services to achieve greater efficiencies in their respective domains.

2-12. Fundamentally, joint interdependence means that each Service depends on the others and on the joint force for key capabilities. It is based on recognition that the Armed Forces fight as one team of joint interagency and multinational partners. Joint interdependence extends combined arms synergy into the joint realm. It is more than interoperability. It is the assurance that Service forces can work together smoothly. It is even more than integration to improve their collective efficiency and effectiveness. Joint interdependence purposefully combines Service capabilities to maximize their complementary and reinforcing effects while minimizing their vulnerabilities. The challenges of the security environment, complexity of unified action, and capabilities required to conduct full spectrum operations make joint interdependence imperative. This is especially true in the area of the GE function where each Service provides both unique and reinforcing capabilities to enhance the efforts of other Service engineer efforts. The Army has a number of unique capabilities not organic to the other Services.

2-13. Unified action for the GE effort is a complex process. The GE function supports the CCDR at the strategic, operational, and tactical levels and must be integrated throughout. Engineers may find themselves providing GE support as part of a multinational, joint, or Army command executing its responsibilities under 10 USC. They may also find themselves supporting and coordinating in an interagency environment, such as with the USAID or numerous NGOs. C2 of engineer units conducting GE functions in such environments is discussed in chapter 3. Planning considerations for GE in the OE are discussed in chapter 4.

This page intentionally left blank.

Chapter 3

Command and Control of General Engineering Operations

Never tell people how to do things. Tell them what to do and they will surprise you with their ingenuity.

General George Patton

C2 of the GE effort is particularly challenging in the OE. Rapid decisive operations, complex joint and multinational environments, simultaneous full spectrum operations, and asymmetric threats are a few of the factors that make effective C2 difficult. Careful consideration during planning and execution must be made in order for C2 of the GE effort to maintain seamless support throughout the OE. This manual will not unnecessarily repeat the overarching doctrine for C2 contained in capstone doctrine, such as FM 3-0, FM 3-34, JP 3-0, or JP 3-34. Nor will it repeat the C2 doctrine in engineer implementing manuals, such as FM 5-116, FM 5-100-15, or FM 5-71-100 (all three of these manuals are scheduled to be incorporated into FM 3-34.23 that will be titled *Engineer Operations – Echelons Above Brigade*). Instead it will discuss the aspects of C2 that impact on GE missions and their implementation.

JOINT COMMAND AND CONTROL

3-1. The JFC will organize his forces so that they are adaptable and tailored to meet specific needs. The JFC establishes his engineer staff to plan and synchronize combat, geospatial, and GE functions in a manner that best supports his mission. In doing so, he may organize his engineer staff in the following three ways:

- As an engineer special staff element that typically reports through the Chief of Staff (COS) to the commander while integrating across all staff sections. This typically occurs when engineer requirements are balanced between combat and sustainment operations. This option provides the greatest flexibility in orchestrating diverse engineer operations and all of the engineer functions, providing the greatest visibility of engineer capabilities, requirements, and responsibilities throughout the staff. The larger the engineer commitment to an operation, the more likely it is to occur.
- As an engineer staff section within the operations directorate (operations staff section [J-3]). This is most likely to occur when the engineer effort focuses on supporting operational movement, fires, and protection.
- As an engineer staff section within the logistics directorate (logistics staff section [J-4]). This is most likely to occur when the engineer effort predominantly supports sustainment operations, or the majority of engineer issues tend to revolve around missions that support logistics operations.

3-2. In each of these staffing options, the joint engineer is likely to assign geospatial engineering assets to the intelligence directorate (intelligence staff section [J-2]).

3-3. The Army engineer assigned as a joint engineer staff officer must be well trained to effectively plan and synchronize joint GE operations as a member of a JFC staff. Action officers must be well versed in the complex C2 of the joint force and the capabilities of all Service engineer forces to maximize their capabilities. Army engineers must recognize that sister Service engineer forces organize and equip their engineer forces for different functions and adapt their capabilities to meet specific needs. This is particularly critical for GE where each Service has capabilities that must be understood to effectively employ them. FM 3-34 and JP 3-34 provide the necessary baseline planning information for all Service engineer capabilities.

3-4. The JFC will organize his forces to most effectively use his resources. He has several options, each of which will have specific implications on how the GE function is executed. These options include—

- **Service component command.** This traditional relationship would have Army engineer forces primarily assigned and under the operational control (OPCON) of the ASCC. The GE effort is directed at providing assured mobility of Army forces from ports of debarkation (PODs) to forward AOs.

- **Functional component command.** Using this relationship, the JFC organizes his forces to accomplish specific functional tasks or missions. Army forces are often part of the joint force land component command (JFLCC) but because GE requirements exist across the joint community, Army engineers will likely support air, maritime, or special operations components as well.

- **Subordinate JTFs.** The JFC may establish one or more JTFs to accomplish focused missions/operations. The JTF will consist of assets from the Services as required to execute specific tasks. The JFC may establish an engineer JTF when the situation dictates that engineer tasks are predominant for a given mission or operation. This is most likely to occur during stability operations or civil support operations where the primary focus is on relieving the suffering from man-made or natural disasters. The Engineer Regiment's 412th and 416th theater engineer commands are examples of organizations that can provide JTF C2 structure. Theater engineer commands should anticipate conducting extensive GE functions in such a situation.

3-5. The GE effort may be greatly impacted by various joint boards that assist the JFC or JTF commander in establishing priorities and policies for the GE effort. Three typical boards may include the Joint Civil-Military Engineering Board, Joint Facilities Utilization Board, and Joint Environmental Management Board (JEMB). Other boards or cells may also be created. Each addresses a separate need within the joint force and is described in detail in JP 3-34. The JFC tailors the scope, roles, and responsibilities of each board to meet the specific operational needs. To adequately synchronize the GE effort, engineers must understand the doctrinal aspects of the board and the theater-specific procedures.

ARMY SERVICE COMPONENT COMMAND AND CONTROL

3-6. Employment of Army engineer assets occurs in a joint environment through C2 exercised by the JFC commander and his supporting engineer staff. The senior Army headquarters under the JTF serves as the ASCC and maintains Army Title 10, USC responsibilities to provide and sustain forces assigned to the AOR. Designated ASCCs, other numbered armies, and corps headquarters can serve as the base for Army forces, joint force land components, and JTFs. Division headquarters may serve as the Army forces to a JTF in small-scale contingencies. Note that an ASCC may simultaneously serve as a JFC or commander of a JTF while still maintaining his responsibilities as an ASCC. Many of the 10 USC responsibilities required of an ASCC are GE-intensive and must be assumed by the senior Army ENCOORD in the AOR. The 10 USC is the codification by subject matter of the general and permanent laws of the United States and is divided into 50 titles. 10 USC is associated with the Armed Forces. The five subtitles of 10 USC include Subtitle A, General Military Law; Subtitle B, Army; Subtitle C, Navy and Marine Corps; Subtitle D, Air Force; and Subtitle E, Reserve Components (see FM 3-34 and JP 3-34 for more information).

3-7. An ENCOORD exists at each echelon to include the JFC, JTF, and ASCC to assist the commander in executing and controlling engineer operations and planning future missions. When serving on a joint staff, he is the joint ENCOORD. The ENCOORD often simultaneously serves as the senior engineer commander of forces employed at that particular echelon. For example, the commander of the engineer brigade may also serve as the ENCOORD for that division or corps headquarters which the engineer brigade is supporting. To facilitate planning and synchronization at echelons above the BCT, the division or corps ENCOORD has a staff focused primarily on synchronizing engineer operations within the higher headquarters plan. ENCOORD responsibilities include—

- Planning and controlling engineer functions.
- Planning and coordinating with the Army component Assistant Chief of Staff, Operations and Plans (G-3)/battalion or brigade operations staff officer (S-3) and fire support coordinator (FSCOORD) on integrating obstacles and fires.
- Advising the commander on using all engineer assets.

- Advising the commander on employing and reducing obstacles.
- Participating in targeting meetings.
- Advising the commander on environmental issues, coordinating with other staff members to determine the impact of operations on the environment, and helping the commander integrate environmental considerations into decision making.
- Providing a terrain visualization mission folder to determine the effects of terrain on friendly and enemy operations.
- Managing the digital terrain data storage device (coordinating with the Army component Assistant Chief of Staff, Intelligence [G-2]/battalion or brigade intelligence staff officer [S-2] for planning and distribution).
- Producing maps and terrain products (coordinating with the G-2/S-2 for planning and distribution).
- Planning and supervising construction, maintenance, and repair of camps and facilities for friendly forces and internees.
- Planning and coordinating using the scatterable mines (with the FSCOORD).
- Providing information on the status of engineer assets on hand.
- Planning and coordinating environmental protection, critical areas, and protection levels.
- Preparing and maintaining the running estimate in assisting the G-2/S-2 with IPB.
- Recommending MSRs and logistic areas, based on technical information to the Army component Assistant Chief of Staff, Logistics (G-4)/battalion or brigade logistics staff officer (S-4).
- Planning the reorganization of engineers to fight as infantry, when the commander deems their emergency employment necessary.
- Coordinating with interagency department engineers, such as the Federal Bureau of Investigation engineer.
- Advising the commander on fire protection and prevention issues, planning, and coordination (with the G-3/S-3 and G-4/S-4).

GENERAL ENGINEERING AT THE THEATER LEVEL

3-8. Theater engineer commands and engineer brigades are the primary building blocks through which the commander provides C2 for the theater strategic and operational level engineer effort. The ASCC who may also serve as the JFC, JFLCC, or as JTF commander determines who will serve as the ENCOORD. It may be the theater engineer command commander, an engineer brigade commander, or other permanently assigned staff engineer (Deputy COS, Engineer). The theater level ENCOORD is responsible for—

- Ensuring that adequate engineer assets are deployed to meet engineer requirements and assure mobility to the theater to support the strategic and operational objectives of the JFLCC and ASCC.
- Integrating all engineer functions into operational plans.
- Task-organizing engineer units to ensure seamless engineer support from the theater to tactical levels.
- Integrating and leveraging USACE capabilities into the operational plan.
- Requesting and integrating joint and multinational engineer assets into the engineer plan. This is particularly important for the GE effort as the U.S. Navy and Air Force have robust GE capabilities.
- Serving on joint engineer and targeting boards as required.
- Coordinating, executing, and managing all GE tasks associated with RSO&I.
- Establishing Class IV requirements and recommending priorities for distribution to the commander.

- Recommending to the commander the establishment of theater engineer work lines (TEWLs) as required and adjusting them throughout the plan.
- Developing and recommending theater level quality of life standards to the commander.
- Ensuring environmental compliance, as far as practicable within the confines of mission accomplishment, with all applicable domestic environmental laws; relevant country-specific final governing standards; the DOD Overseas Environmental Baseline Guidance Document; relevant international agreements; "Environmental Considerations" annexes or appendixes to the relevant operation plan (OPLAN), operation order (OPORD), and/or other operational directives; and any other environmental requirements that apply to the operation (see the Judge Advocate General for additional questions).

3-9. Due to the complexity and variety of operations at theater level, there is no specific task organization that can be used under all situations. The ENCOORD must use a modular system to build the engineer organizations required to meet theater needs. Modularizing the force required to conduct GE missions leads to a more effective joint and expeditionary mind-set and has some of the following attributes:

- Scalable.
- Tailorable C2.
- Projectable capabilities.
- Reduced footprints.
- Functional incremental employment.
- Modular design of all systems and organizations.

3-10. This will ensure that the right mix of force structure is deployed and minimize the footprint of deployed engineer forces. See FM 5-71.100 or FM 5-116 for more detail.

GENERAL ENGINEERING AT OPERATIONAL AND TACTICAL LEVELS

3-11. The corps (and potentially the division) headquarters is usually the primary organization responsible for translating theater level operational objectives into tactical objectives. Engineers assist in achieving these tactical objectives by supporting the corps and division commander's assured mobility needs at the operational and tactical levels. The ENCOORD at these echelons are dual-hatted as an engineer brigade commander. As such, the engineer brigade commander is responsible for task-organizing all engineers within the corps or division AO to achieve operational and tactical GE effects. This includes engineer resources that support the BCTs.

COMMAND AND SUPPORT RELATIONSHIPS

3-12. One of the most critical aspects of ensuring adequate GE support is assigning the proper command and support relationship to subordinate units (Table 3-1, page 3-6). Further guidance for establishing theater engineer command and support relationships is available in FM 3-0, FM 6-0, FM 3-34, and JP 3-34. ENCOORDs must carefully examine the required GE effort when recommending command and support relationships. Some specific considerations include—

- The GE effort will often include units that have specialized or functional capabilities. Examples of these types of modular units include dive teams, firefighting teams, prime power units, and paving teams. ENCOORDs must ensure that these assets are available at the needed time and place and ensure that they are provided with the necessary support to complement their very limited organic support. These units are usually provided with a C2 relationship to a larger engineer headquarters.
- Units with GE capabilities often have low-density equipment with high maintenance requirements. Requisition of repair parts is often difficult and the number of personnel trained on maintenance of this sort of equipment is often limited.

- GE tasks often require large amounts of supply, such as lumber, gravel, concrete, and asphalt. These can put a large strain on the supply system and units must have the capability to procure such materials locally if possible.
- The majority of the Engineer Regiment's GE capability resides in the reserve components. This means that the time necessary to prepare these units for employment may be longer than if they were active Army units.
- GE missions are often time intensive. Rapidly changing tactical conditions can affect the priority of effort and affect the completion of GE missions.
- Units with the required specific capabilities will almost always be in short supply. ENCOORDs must carefully prioritize their efforts, then establish command and support relationships that maximize the effectiveness of these units with specialized capabilities.
- GE tasks occur throughout the OE from strategic through tactical levels. Different echelons have very different needs for the type of GE support required and different abilities to logistically support GE units.
- The type of operation (offense, defense, stability, or civil support) that is being executed will greatly influence where in the AO and at what echelon GE tasks tend to be accomplished.
- The engineer work line (EWL) can be used as a means of describing unit responsibilities for GE tasks. This must be tied to the command and support relationship being assigned (table 3-1, page 3-6).

Table 3-1. Command and support relationships

IF RELATIONSHIP IS—		Inherent Responsibilities							
		Has Command Relationship with the—	May be Task-Organized by the—	Receives logistical support and sustainment from the —	Assigned Position or AO by the—	Provides Liaison	Establishes/ Maintains Communications	Has Priorities Established by—	Gaining Unit Can Impose Further Command or Support Relationship of—
COMMAND	Attached	Gaining unit	Gaining unit	Gaining unit	Gaining unit	As required by gaining unit	With unit to which attached	Gaining unit	Attached: OPCON; TACON; GS; GSR; R; DS
	OPCON	Gaining unit	Parent unit and gaining unit; gaining unit may pass OPCON to lower headquarters. See Note 1	Parent unit	Gaining unit	As required by gaining unit	As required by gaining unit and parent unit	Gaining unit	OPCON; TACON; GS; GSR; R; DS
	TACON	Gaining unit	Parent unit	Parent unit	Gaining unit	As required by gaining unit	As required by gaining unit and parent unit	Gaining unit	GS; GSR; R; DS
	Assigned	Parent unit	Parent unit	Parent unit	Gaining unit	As required by parent unit	As required by parent unit	Parent unit	Not applicable
SUPPORT	DS	Parent unit	Parent unit	Parent unit	Supported unit	Supported unit	With parent unit; supported unit	Supported unit	See Note 2
	Reinforcing (R)	Parent unit	Parent unit	Parent unit	Reinforced unit	Reinforced unit	With parent unit; reinforced unit	Reinforced unit; then parent unit	Not applicable
	GSR	Parent unit	Parent unit	Parent unit	Parent unit	Reinforced unit and as required by parent unit	With reinforced unit and as required by parent unit	Parent unit; then reinforced unit	Not applicable
	GS	Parent unit	Parent unit	Parent unit	Parent unit	As required by parent unit	As required by parent unit	Parent unit	Not applicable

Notes.

1. In NATO, the gaining unit may not task-organize a multinational unit (see TACON).

2. Commanders of units in DS may further assign support relationships between their subordinate units and elements of the supported unit after coordination with the supported commander.

ENGINEER WORK LINE

3-13. The *EWL* is a coordinated boundary or phase line used to compartmentalize an AO to indicate where specific engineer units have primary responsibility for the engineer effort (the definition was shortened, and the complete definition is printed in the glossary) (see FM 1-02 and FM 3-34). The EWL may also be used as a boundary between engineer organizations but this should not be its primary purpose. It may or may not follow maneuver unit boundaries. Traditionally, it is used at the division level to discriminate between engineer assets assigned to the division level and higher echelon engineer units. It also serves as a visualization tool for the ENCOORD for where the GE effort primarily occurs. Forward of the EWL is focused on combat engineering functions and tasks with minimal GE being accomplished. It is behind the EWL where the majority of resource-intensive GE tasks are performed.

3-14. Use of an EWL as a visualization tool and C2 measure as depicted in figure 3-1 is effective on the contiguous battlefield. Use of this C2 measure in such a manner allows engineer units organic and augmenting BCTs to focus on providing robust combat engineering and limited GE support forward of the EWL. To the rear of the EWL, uncommitted echelons above BCT engineer units in a DS or GS role to the division focus primarily on GE tasks that are part of the sustainment of the division.

Such tasks may include MSR upgrade and repair, facilities construction, repair of field landing strips, LOC bridging, and other sustainment of the force.

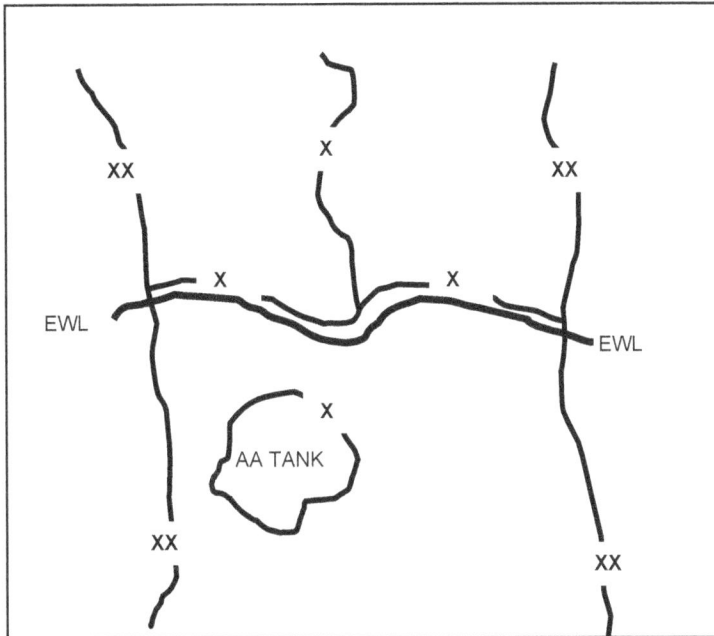

Figure 3-1. Division EWL in contiguous operations

3-15. Use of an EWL as a C2 measure must be carefully applied in the noncontiguous AO. Figure 3-2, page 3-8, depicts an example of multiple EWLs to depict responsibilities between organic and augmenting engineers to the BCTs and echelons above BCT engineer units. In this case, organic and augmenting BCT engineers would focus on engineering tasks inside EWLs Dog, Cat, and Lion, while echelon above BCT engineers would be responsible for GE tasks throughout the remainder of the division AO, to include the intermediate staging base (ISB). This construct may be useful to the ENCOORD during the offense and defense where the focus is providing support to the BCTs and combat engineering support to combat maneuver forces. However, during stability operations or civil support operations, GE tasks will be executed with echelons above BCT engineers operating throughout the AO.

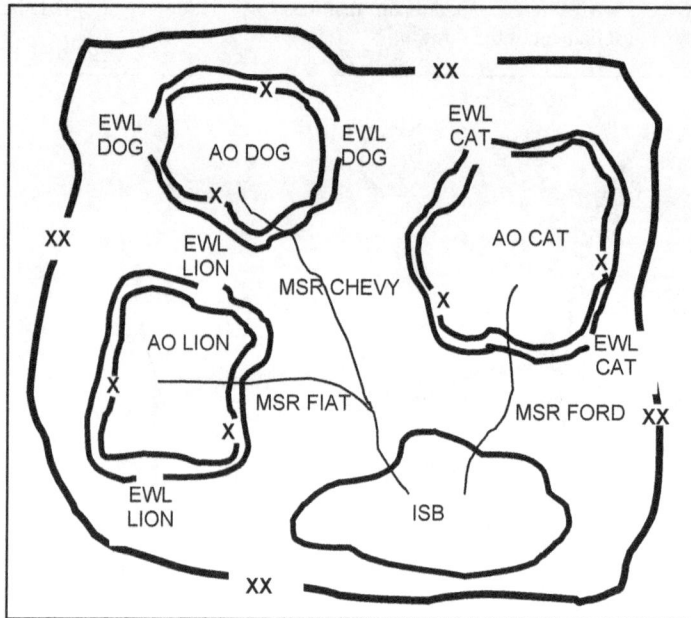

Figure 3-2. Division EWL in noncontiguous operations

3-16. At corps and theater levels, the ENCOORD for that echelon may establish a corps engineer work line (CEWL) or TEWL in much the same manner as the division ENCOORD. The theater ENCOORD augments subordinate echelons by assuming responsibility for specific support on a task basis forward to the appropriate EWL, thus releasing the DS and GS GE units to engage in activities as far forward as possible.

3-17. The ENCOORD that assigns the EWL to a particular area of operations is responsible for planning and advising his commander on when to shift its location. This occurs after a careful analysis of the ongoing operation, available GE assets, and future requirements. Early in a contingency, it may be very difficult for the theater ENCOORD to shift the TEWL out of the theater staging base because of shortages in GE assets.

Chapter 4

Planning Considerations and Tools

Seek first to understand, then to be understood.

Stephen R. Covey

Planning is the means by which the commander envisions a desired outcome, lays out effective ways of achieving it, and communicates to his subordinates his vision, intentions, and decisions, focusing on the results he expects to achieve. During the conduct of the planning, commanders and staffs must understand that the types and magnitude of GE tasks that can be accomplished will vary based on the type, capability, and resources available to the engineer unit assigned the mission. It is during the continuous refinement of the mission analysis in the military decision-making process (MDMP) and the engineer estimate where it is determined if the proper assets are available to accomplish all the tasks. The MDMP process outlined in FM 5-0 is as sound for executing the GE function as it is for any other military mission. There are particular considerations and tools for planning GE missions that must be understood and integrated into the process to make them an effective portion of the planning process. The intent of this chapter is to highlight these considerations and tools.

MILITARY DECISION-MAKING PROCESS

4-1. The MDMP is well defined in FM 5-0. JP 5-0 provides the planning construct in a joint environment in much the same manner. Many GE tasks have unique requirements that must be considered when applying the MDMP to a specific mission. Table 4-1, page 4-2, lists some of the potentially unique aspects of GE missions as they pertain to each step of the MDMP. Although not nearly all-inclusive of the variations required of the GE effort, this list demonstrates that GE missions and tasks are often complex, resource intensive, and require extensive and proactive coordination. Additionally, a successful GE effort requires an understanding of all engineer requirements (combat, general, and geospatial) and their roles in the concept of operations.

Table 4-1. GE in the MDMP

Steps of the MDMP	GE Considerations
Receipt of the Mission	Review the joint operational order, particularly appendix 8, annex C; appendix 15, annex C; appendix 16, annex C; appendix 5, annex D; appendix 6, annex D; annex G; and annex L.
	Receive higher headquarters construction directives.
	Request geospatial and medical information about the AO.
	Establish several potential engineer-related boards.
Mission Analysis	Determine availability of construction materials.
	Review the availability of construction assets to include Army, joint, multinational, HN, and contract.
	Determine and review theater construction standards and base camp master planning documentation if required.
	Review UFC as required.
	Review existing geospatial and medical data on potential sites, conduct site reconnaissance (if possible), and determine the threat (to include environmental and EH).
	Obtain the necessary geologic, hydrologic, and climatic data.
	Determine the level of interagency cooperation required.
	Determine funding sources as required.
COA Development	Produce different designs that meet the commander's intent (use the TCMS when the project is of sufficient size and scope).
	Determine alternate construction locations, methods, means, materials, and timelines to give the commander options.
	Determine the real property and real estate requirements.
COA Analysis	Use the critical path method to determine the length of the different COAs and the ability to crash the project.
COA Comparison	Determine the most feasible, acceptable, and suitable methods of completing the GE effort.
	Determine and compare the risks of each GE COA.
COA Approval	Gain approval of the construction management plan, safety plan, security plan, logistics plan, and environmental plan as required.
Orders Production	Produce construction directives as required.
	Provide input to the appropriate plans and orders.
	Ensure that all resources are properly allocated.
Rehearsal	Conduct construction prebriefings.
	Conduct preinspections and construction meetings.
	Synchronize the construction plan with local and adjacent units.
Execution and Assessment	Conduct quality assurance and mid-project inspections.
	Participate in engineer-related boards.
	Maintain "as-built" and "red-line" drawings.
	Project turnover activities.

JOINT GENERAL ENGINEERING PLANNING CONSIDERATIONS

4-2. The primary joint doctrinal publication for GE operations is JP 3-34. Army planners must understand, however, that the Air Force and Navy have a narrower focus for the GE mission and often refer to it as *civil engineering* (those sustainment activities that identify, design, construct, lease, or provide facilities and which operate, maintain, and perform war damage repair and other engineering functions in support of military operations [JP 1-02]). The Air Force and Navy consider general (civil) engineering to be primarily a logistics function that is executed to sustain their forces in a contingency operation. Their activities tend to focus on missions such as base camp and life support development, construction and repair of seaports of debarkation (SPODs), aerial ports of embarkation (APOEs), and other facilities and sites, and not focus on operational support to ground maneuver forces. GE contains an operational support piece that, while different from combat engineering, does provide support to maneuver commanders and facilitates their assured mobility, although this is typically at the operational level rather than the tactical level.

4-3. The Engineer Support Plan (appendix 6, annex D) is produced by a joint staff for input to a joint OPLAN as part of the deliberate planning process. It ensures that essential general (civil) engineering capabilities are identified and will be provided at the required locations and times. It is the most critical appendix for GE in a joint OPLAN. Other critical portions of a joint OPLAN for GE planning include—

- Appendix 8, annex C: Air Base Operability.
- Appendix 15, annex C: Force Protection.
- Appendix 16, annex C: Critical Infrastructure Protection.
- Appendix 5, annex D: Mobility and Transportation.
- Annex G: CA.
- Annex L: Environmental Considerations.

UNIFIED FACILITIES CRITERIA

4-4. Unified Facilities Criteria (UFC) provides facility planning, design, construction, operations, and maintenance criteria for all DOD components. Individual UFC are developed by a single-discipline working group and published after careful coordination. They are jointly developed and managed by the USACE, the NAVFAC, and the Air Force Civil Engineering Support Agency. Although UFC are written with long-term standards in mind, planners who are executing under contingency and enduring standards for GE tasks will find them useful. Topics include pavement design, water supply systems, military airfields, concrete design and repair, plumbing, electrical systems, and among others.

4-5. UFC are living documents and will be periodically reviewed, updated, and made available to users as part of the Services' responsibility for providing technical criteria for military construction. UFC are effective upon issuance and are distributed only in electronic media from the following sources:

- UFC Index <http://65.204.17.188//report/doc_ufc.html>.
- USACE TECHINFO <http://www.hnd.usace.army.mil/techinfo/index.html>.
- NAVFAC Engineering Innovation and Criteria Office <http://criteria.navfac.navy.mil>.
- Construction Criteria Base (CCB) system maintained by the National Institute of Building Sciences (NIBS) at <http://www.nibs.org/ccb>.

4-6. GE planners must consider all construction standards established by CCDRs and ASCCs for their AOR. Specific examples of these are the United States European Command (USEUCOM) Red Book AOR and United States Central Command (USCENTCOM) Sand Book. These constantly evolving guidebooks specifically establish base camp standards that consider regional requirements for troop living conditions and, therefore, have a major impact on projects, such as base camps and utilities. Because availability of construction materials may vary greatly in various AORs, standards of construction may differ greatly between them. CCDRs often establish standards for construction in OPORDs and fragmentary orders (FRAGOs) that may take precedence over guidebooks. Planners must understand the expected life cycle of a GE project to apply these standards. Often the standards will be markedly different, depending on whether the construction is contingent or is intended to have an enduring presence.

OPERATIONAL AND TACTICAL PLANNING CONSIDERATIONS

PROJECT MANAGEMENT

4-7.   Planners use the project management system described in FM 5-412 as a tool for the process of coordinating the skill and labor of personnel using machines and materials to form the materials into a desired structure. Figure 4-1 shows the project management process that divides the effort into preliminary planning, detailed planning, and project execution. Today, when engineer planners are focused on GE tasks, they rely extensively on the Theater Construction Management System (TCMS) to produce the products required by the project management system. These products include the design, activities list, logic network, critical path method or Gantt chart, bill of materials (BOM), and other products. Effective products produced during the planning phases also greatly assist during the construction phase. In addition to TCMS, the engineer has various other reachback tools or organizations that can exploit resources, capabilities, and expertise that is not organic to the unit that requires them. These tools and organizations include, but are not limited to, the USAES; USACE Engineering Infrastructure and Intelligence Reachback Center (EI2RC), and the Tele-Engineering Operations Center (TEOC), Engineering Research and Development Center (ERDC), 412th and 416th Theater Engineer Commands; the Air Force Civil Engineer Supporting Agency; and the NAVFAC (see appendix B).

4-8.   The project management process normally begins at the unit level with the construction directive. This gives the who, what, when, where, and why of a particular project and is similar to an OPORD in its scope and purpose. Critical to the construction directive are plans, specifications, and all items essential for success of the project. Units may also receive GE missions as part of an OPORD, FRAGO, warning order (WARNORD), or he may receive them verbally. When a leader analyzes a construction directive, he may need to treat it as a FRAGO in that much of the information required for a thorough mission analysis may exist in an OPORD issued for a specific contingency operation.

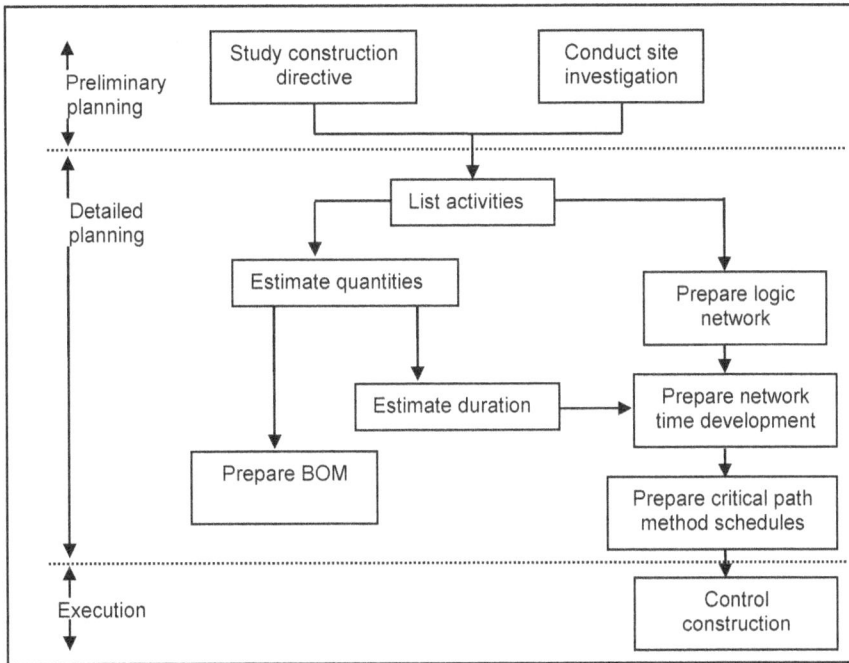

Figure 4-1. Project management process

INFRASTUCTURE RECONNAISSANCE: ASSESSMENT AND SURVEY

4-9.   Transitions between offensive, defensive, stability, and/or civil support operations are a complex process. Engineers conducting offensive operations one day may suddenly find themselves conducting stability operations. Likewise, stability operations may suddenly turn violent and units will find themselves on the defense in preparation for offensive operations. The keys to success under such circumstances are adaptable, flexible small-unit leaders who are able to rapidly assess the situation and make decisions with minimal guidance and intent.

4-10. Recent history in Operation Just Cause (Panama), Operation Desert Storm, Operation Enduring Freedom, and Operation Iraqi Freedom have been noted as relatively short and violent offensive operations followed by a rapid transition to stability operations. Engineer units focused on the mobility of the force suddenly found themselves relieving the suffering of the local population, restoring infrastructure, and providing basic services (including general security). Inherent in this transition is a shift in focus from combat to the GE function. To successfully make this transition, leaders must determine the requirements inherent in this transition.

4-11. The infrastructure assessment and the infrastructure survey are two types or levels of reconnaissance used to gather this necessary infrastructure information. The purpose of the assessment is to provide immediate feedback concerning the status of the basic services necessary to sustain the local population. The memory aid to describe this assessment is sewage, water, electricity, academics, trash, medical, safety, and other considerations (SWEAT-MSO) with each of the letters describing a major area within the assessment (figure 4-2). The model can be adapted for use at the tactical level in either stability or civil support operations. In either type of operation, the SWEAT-MSO model can be used during COA development to delineate tasks, missions, and effects that support civil-military related objectives.

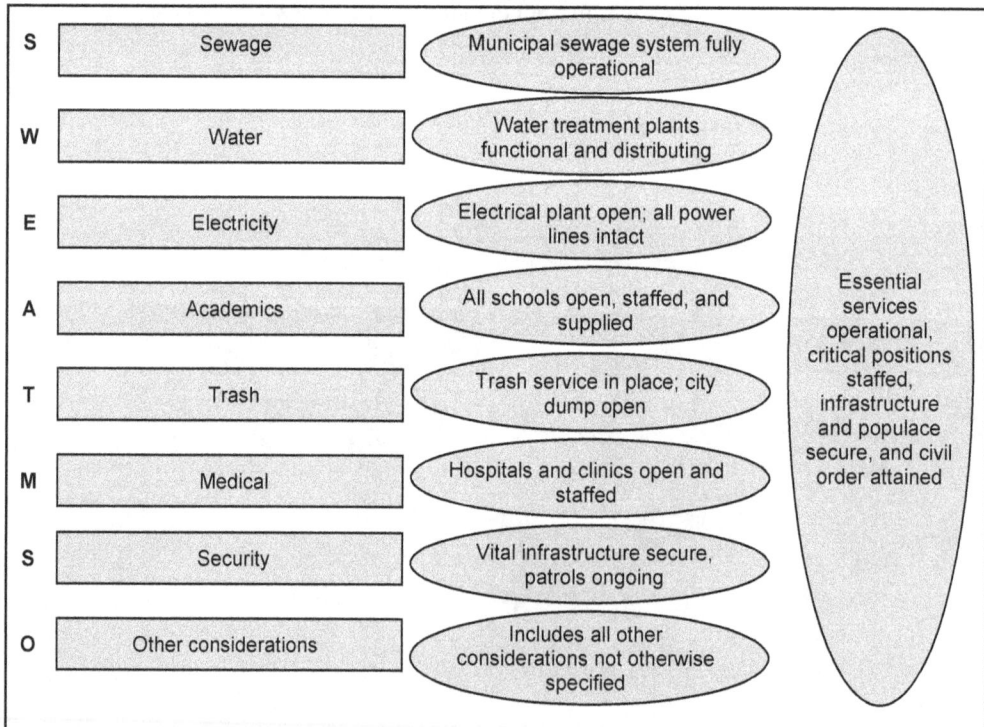

Figure 4-2. The infrastructure assessment and survey model

4-12. The basic services or categories evaluated depend on the situation, mission, and the commander's intent. While it is typically performed by engineers, it may be accomplished by others when an engineer is not available, depending on the expertise available and the desired type or quality of information required. If available, leaders should also consult military and NGO units and agencies in the area to determine if there are extenuating circumstances that may influence the outcome of the assessment. Typically, engineer planners use this information to define immediate needs and determine priorities of work. While an infrastructure assessment is designed to support the resolution of the immediate challenges, it will set the conditions for a successful transition. Leaders must continue to expand and refine the assessment. As follow-on to the assessment, the infrastructure survey provides a very detailed description of the condition of major services. The primary difference in the assessment and the survey is the degree of technical information and the expertise required. The survey is conducted by forward USACE personnel assigned to a forward engineer support team (FEST) and will integrate other technical specialties (medical, CA, and others) to enhance the quality of the survey.

4-13. Some of the primary considerations for the assessment are—

- **Sewage.** Determine what the status of the local sewage system is. Determine what health and environmental risks exist.
- **Water.** Determine what potable water sources are available. Determine if they are adequate and if they have been tested.
- **Electricity.** Determine the status of electrical generation facilities, to include the availability of generators. Determine the status of the transmission infrastructure. Determine what critical facilities (to include hospitals, government buildings, and schools) are not having their needs met. Determine the availability of fuel for transportation, heating, and cooking. Determine if there is an adequate system of distribution.
- **Academics.** Determine what schools are in need of repair and rebuilding.
- **Trash.** Determine if there is a system in place for the removal of waste. Determine what hazardous waste streams are being generated that may have detrimental impacts on health and the environment. Determine the ultimate disposal system for trash.
- **Medical.** Determine the medical services available and if they are operational. Determine if an emergency service exists. Determine if there are services available for animals.
- **Safety.** Determine if there are police and fire services available. Determine if unexploded explosive ordnance (UXO) or other EH is an issue.
- **Other considerations.** Other considerations that leaders may consider as a part of the assessment include—
    - **Transportation networks.** Determine if there are roads, bridges, and railroads that are trafficable. Determine if there is an operational airport and if there are usable helicopter landing sites. Determine if they can sustain the local, humanitarian assistance traffic.
    - **Fuel distribution.** Determine if there is a fuel distribution system available to commercial and residential customers.
    - **Housing.** Determine if homes are structurally sound and habitable and if they include basic utilities.
    - **EH.** Determine if there are any EH.
    - **Environmental hazards.** Determine if there are any environmental hazards.
    - **Communications.** Determine if there is an operational telephone network available. Determine if the town has television, radio, and newspaper access and, if so, do they work.
    - **Places of worship.** Determine if there are adequate facilities to support religious activities for all groups.
    - **Attitude.** Determine if local people and community leaders are supportive and if there is ethnic tension.

4-14. Table 4-2, page 4-8 through 4-10, provides an example checklist for an initial infrastructure assessment of a town or location to assist in determining the HCA needs of the town at the beginning of a stability and reconstruction operation. This example is not intended to be all-inclusive, but rather another aid to support an assessment. The formal survey will be much more specific and in-depth than the information in any assessment. The example uses the SWEAT-MSO model introduced in Figure 4-2, page 4-6, to assist the Soldier in organizing an assessment strategy. Appendix C provides an infrastructure assessment rating to assist in rating each of the assessed categories. Leaders may use these resources to begin developing priorities, obtaining resources, and refining a plan. Many of the tasks derived from this process will be GE tasks (such as facilities construction, well drilling, power generation, and road repair).

4-15. Leaders should understand that many GE efforts will be part of larger information operations (IO). GE tasks for infrastructure repair must be coordinated through the humanitarian assistance coordination center (HACC), the civil-military operations center (CMOC), and possibly the fire support coordinator and fire support cell, to achieve proper synchronization and the effect desired. Engineers will integrate expertise (CA, medical, psychological operations [PSYOP], and others) during their reconnaissance and project work, including coordinating with the maneuver commander in charge of the specific AO where a project is located. Leaders are encouraged to modify the contents of table 4-2, pages 4-8 through 4-10, to meet the specific needs and requirements of the operation and discuss its contents with all personnel involved before undertaking a reconnaissance. Leaders should reference FM 3-05.40 and FM 3-13 for more information on conducting these operations as part of a combined arms team. A more in-depth discussion of infrastructure reconnaissance and how to conduct it as part of a combined arms team is included in FM 3-34.170.

Table 4-2. Sample infrastructure assessment

GE Requirements – Infrastructure Assessment			
Town/Village/Neighborhood:	Location:		Assessor:
Local Points of Contact (Name, Location, Telephone Number, and so forth)			
Mayor: City Council: City Engineer: Religious Leaders: Community Leaders:		Police Chief: Fire Chief: School Administrators: NGOs:	
Population:	Male:	Female:	Religious Breakdown:	Ethnic Breakdown:

S	Sewage System Assessment

Status of municipal sewage system and distribution system:

Status of sewage systems in commercial and residential properties:

Immediate needs:

W	Water Assessment

Status of water treatment plants and distribution systems:

Status of potable water in commercial and residential properties:

Storage capacity:

Wells (location and capacity):

Immediate needs:

E	Electricity Assessment

Status of electric plant and distribution system:

Status of electric power in commercial and residential properties:

Alternate power sources:

Immediate needs:

Table 4-2. Sample infrastructure assessment (continued)

A	Academics Assessment
Status of school buildings: Status of teachers and supplies: Immediate needs:	

T	Trash Assessment
Status of trash collection system: Status of disposal site: Immediate needs:	

M	Medical Assessment
Status of hospital and clinic buildings: Status of physicians and supplies: Immediate needs:	

S	Safety Assessment
Status of police and fire departments: Status of safety personnel and supplies: Immediate needs:	

O	Other Considerations
Transportation System Assessment	
Status of road system (attach sketch if necessary): Impact on critical transportation needs: Immediate needs:	
Fuel Distribution Assessment	
Status of fuel distribution system: Storage capacity: Immediate needs:	

Table 4-2. Sample infrastructure assessment (continued)

Housing Assessment
Status of structures: Status of utilities: Immediate needs:
EH Assessment
Explosive ordnance locations and type (send 9-line UXO report as required by the mission): Explosive ordnance marked (if yes, marking description): Immediate needs:
Environmental Hazards Assessment
Do known hazards exist (if yes, describe): Are chemicals visible on the ground (if yes, describe): Abandoned manufacturing buildings (if yes, are waste products and streams contained?): Immediate needs:
Other Critical Considerations:
Recommended Priorities:
Remarks:

Signature:	Organization:	Date:

FIELD FORCE ENGINEERING

4-16. The overarching concept of FFE is provided in FM 3-34. It is the application of all the Engineer Regiment's capabilities across full spectrum operations facilitated by both forward presence and reachback (see appendix B). FFE works to provide seamless specialized GE support to any type of military operation, including military support to the Department of Homeland Security (DHS) during civil disaster response. FFE fuses the capabilities resident in USACE, USAES, theater engineer commands, Public Works, and civilian contractors. It recognizes the critical need for early, integrated engineer participation in planning and optimizing engineer capabilities for mission analysis, development, and accomplishment. Although FFE may apply to all engineer functions, GE missions and geospatial engineering support best suit its applications.

FORWARD PRESENCE PLANNING OPTIONS

4-17. The cornerstone of forward presence is the ability to form modular and scaleable teams capable of deploying into theater on short notice to provide engineering support to the CCDR. Engineer planners at all levels must carefully analyze the mission to determine the required level of forward presence support and tailor its requests. Because these teams can be tailored, specificity of requests in terms of the type of missions the team will conduct is critical. To facilitate the engineer planning effort, USACE maintains established liaison officer (LNO) planners at the combatant command (command authority) (COCOM) and ASCC levels.

4-18. The request for forces (RFF) process is the surest means to acquire the services of a deployable USACE organization (described below). Because USACE is a reimbursable organization, the deployment order (DEPORD) produced as a result of the RFF ensures funding for services provided. Requests for USACE support should be channeled through the USACE LNO at the combatant command or ASCC echelons. The USACE Deputy G-3 will respond to requests for engineer support in the event that coordination through the LNOs is not possible.

Forward Engineer Support Team–Advance

4-19. The forward engineer support team-advance (FEST-A) is a deployable planning and assessment cell that augments the engineer or civil military staffs of other organizations from the combatant command down to the BCT level. Normally five to eight personnel, the FEST-A consists of a military team leader; geographic information system (GIS) specialist; and civil, mechanical, and electrical engineers. It can be augmented with structural, environmental, and other engineering skills depending on the mission. A FEST-A prepares plans and designs to include master planning for base camps, I/R facilities, forward operating bases (FOBs), convoy support centers, and other facilities. FEST-A units provide assessments of infrastructure and rough order of magnitude estimates for reconstruction of damaged infrastructure or improvement of existing infrastructure. A FEST-A is also capable of providing formal infrastructure reconnaissance surveys. They deploy with tele-engineering capability and are the focal point for reachback for technical engineering support for the supported unit. During Phase IV of Operation Iraqi Freedom, as many as thirteen FEST-A organizations were deployed in the USCENTCOM AOR at one time in support of military units, the Coalition Provisional Authority, and the USAID. FEST-A units have conducted hundreds of engineer assessments in Iraq and Afghanistan.

Note. T*ele-engineering* assists engineers and the commanders they support in planning and executing their missions with capabilities inherent in field force engineering (FFE) through exploitation of the Army's command, control, and communications architectures to provide a linkage between engineers and the appropriate nondeployed subject matter experts (SMEs) for resolution of technical challenges. Tele-engineering is under the proponency of the USACE.

Forward Engineer Support Team–Main

4-20. The forward engineer support team-main (FEST-M) is a deployable, USACE organization that executes the USACE mission in theater, specifically execution of contract construction. Recently deployed FEST-M units consisted of 12 to 150 personnel depending on the mission. A FEST-M is typically placed under the OPCON of the CCDR, ASCC, or theater engineer command commander. As the most robust of the forward presence organizations, the FEST-M may serve as the nucleus for a precursor to an overseas USACE district. The FEST-M typically has personnel with design capabilities for all disciplines, to include electrical, mechanical, civil, and environmental engineering. The commander of the FEST-M tailors the skill sets of the team to meet mission requirements.

Contingency Real Estate Support Team

4-21. The contingency real estate support team (CREST) is a 2- to 4-person team that is often attached to a FEST-M. A CREST works on a delegation of authority to a acquire real estate for the DA. The CREST can quickly execute real property leases in forward locations. Funding for lease payments must be provided by the supported command. Their support is critical when deploying forces in a permissive environment in countries with a sovereign government. Without this capability, the JFC will not be able to use HN facilities or land.

Environmental Support Team

4-22. The environmental support team is a team of environmental experts established at the USACE division level. It is available to deploy to perform environmental analysis and address environmental issues regarding base camp development and operations. This team is capable of performing an Overseas EBS as directed in DOD Instruction 4715.5, when supported with medical expertise. An EBS is most effective when it is conducted in conjunction with an EHSA.

Water Resource Detection Team

4-23. The water resource detection team identifies high-potential areas for quality water that is within the capability of available drilling equipment. While not automatically deployable, the team provides technical support for military operations using databases, imagery, and other sources. See appendix B for more information.

TECHNICAL ENGINEERING REACHBACK

Engineer Infrastructure and Intelligence Reachback Center

4-24. The Engineer Infrastructure and Intelligence Reachback Center (previously known as the infrastructure assessment team [IAT]) it serves as the USACE FFE "hub" for engineering support and GIS infrastructure intelligence to military deployments and civil-military operations worldwide. It is known as the "one door" to USACE for reachback technical assistance and engineering support. It provides infrastructure assessments, base camp planning, and design assistance in support to the USCENTCOM, USEUCOM, USNORTHCOM/Federal Emergency Management Agency (FEMA), United States Pacific Command (USPACOM), and the United States Southern Command (USSOUTHCOM). See appendix B for more information.

Base Development Team

4-25. The base development team (BDT) provides installation-level base development planning and facilities design expertise for ISBs, base camps, FOBs, and displaced persons camps. It integrates environmental aspects into the design of these facilities. This ten-person nondeployable organization is located in various USACE engineer districts and draws support from the ERDC and other USACE centers of expertise. It is capable of completing a 30 percent design of a major base camp within three days. It uses the inherent capabilities of Army Facilities Components System (AFCS) and TCMS to prepare designs and passes them to forward presence organizations via tele-engineering or the SECRET Internet Protocol

Router Network (SIPRNET). The BDT provided reachback support for FEMA housing planning and response teams for temporary housing during Hurricane Katrina and is prepared to provide support for civil disaster response as needed.

Tele-Engineering Operations Center

4-26. The TEOC serves as the focal point for video conferences. It has a classified bridge which has about 50 ports for video conferences. It also has an unclassified bridge with about 15 ports. See appendix B for more information.

REACHBACK CAPABILITY

4-27. USACE (and other engineer organizations) elements may use the tele-engineering communications equipment-deployable (TCE-D) for reachback for technical engineering support. The TCE-D provides reachback capability using commercial off-the-shelf (COTS) communications equipment with encryption added. Video teleconferences and data transfers can be conducted from remote sites where other means of communication are nonexistent or unavailable. TCE-D comes with its own satellite links and does not use bandwidth from units deployed in theater. Although originally designed for USACE organizations, TCE-D is fielded to a number of tactical units for reachback and has proved to be a valuable tool. See appendix B for more information on the reachback process, organizations, and tools.

PART TWO

Lines of Communication

Part two of this manual discusses specific GE support for the creation and repair of LOCs. LOCs are routes over land, water, or air that connect an operating military force with a base of operations and along which supplies and military forces move. There are strategic, operational, and tactical LOCs and usually establishing and maintaining them are sustainment operations. The joint force depends on ports and airfields for AOR access and links to the CONUS base of operations. They may also have important impacts at the operational and tactical levels. The joint force depends on roads and railways for a link to its base of operations, and bridging is often included to establish them. Engineers support each of these aspects of LOCs and each are addressed here as part of the GE function.

Chapter 5

Seaports of Debarkation

They must float up and down with the tide. The anchor problem must be mastered. Let me have the best solution worked out. Don't argue the matter. The difficulties will argue for themselves.

Winston Churchill on pier construction to support the invasion, May 1943

Obtaining adequate port facilities early in any contingency is essential to the efficient flow of troops and materiel. Port construction, rehabilitation, and repair are of vital importance to the success of any such mission as they support assured mobility at the strategic level. They are most often inherently joint operations. Securing these facilities is often an initial objective of overseas operations. HN agreements granting the military use rights are essential to ensure that the impact on commercial shipping and local military operations is kept to a minimum.

While the situation dictates the COA, assault landing facilities are usually used for supply and replenishment in the initial phase of a campaign, followed by logistics over-the-shore (LOTS) and joint logistics over-the-shore (JLOTS) operations, as discussed later in this chapter. As established port areas are acquired or rehabilitated, LOTS sites are normally abandoned. Certain AOs, however, may require the use of beach sites for extended periods of time or even indefinitely, due to the lack of existing facilities, the geography, the terrain, or the enemy situation. The construction of new ports is normally undesirable, as it requires a large amount of labor, materials, and time, and probably would lack the desirable related facilities, such as connecting road and rail networks. Therefore, existing ports are usually targeted for rehabilitation and upgrade. The engineer mission is to support construction, maintenance, and repair of a wide variety of facilities, both above and below the waterline.

SCOPE OF PORT OPERATIONS

5-1. This chapter is a guide for the construction and rehabilitation of ship-unloading and cargo-handling facilities in the OE. The coverage includes special problems encountered in port construction and the construction of those supporting structures located in and around the port facility. Based on current trends in the commercial shipping industry, it is anticipated that up to 90 percent of all cargo arriving in future OEs will be containerized.

5-2. This method of shipping requires dock and road surfaces capable of withstanding severe loads and heavy lift equipment capable of transferring the largest loaded container (8 feet wide, 40 feet long, and 67,200 pounds) from large, oceangoing vessels to shore facilities. These factors should be considered during port planning. The guidelines concerning facilities for handling containerized cargo and container shipping outlined within this chapter represent the most current developments in this industry.

Perspective

One of the singular logistical achievements of World War II associated with the Normandy invasion was the gigantic artificial harbors, or "Mulberries," that were designed, built, and transported to the landing beaches, which lacked the natural harbor facilities that would be vital to continued support of the invasion. One harbor, known as Mulberry A, was constructed off Saint-Laurent at Omaha Beach in the American sector, and the other, Mulberry B, was built off Arromanches at Gold Beach in the British sector. Each harbor, when fully operational, had the capacity to move 7,000 tons of vehicles and supplies per day from ship to shore. Each Mulberry harbor consisted of roughly 6 miles of flexible steel roadways (code-named Whales) that floated on steel or concrete pontoons (called Beetles). The roadways, which terminated at great pier heads called Spuds, were jacked up and down on legs which rested on the seafloor. These structures were to be sheltered from the sea by lines of massive sunken caissons (called Phoenixes), lines of scuttled ships (called Gooseberries), and a line of floating breakwaters (called Bombardons). It was estimated that construction of the caissons alone required 330,000 cubic yards of concrete, 31,000 tons of steel, and 1.5 million yards of steel shuttering. The various parts of the Mulberries were fabricated in secrecy in Britain and floated into position immediately after D-Day. Within 12 days of the landing (D-Day plus 12), both harbors were operational.

RESPONSIBILITIES AND CAPABILITIES

5-3. The operation of a port in an OE is a large and vital undertaking with many divisions of responsibility between the Navy and the branches of the Army. The geographic CCDR or ASCC makes basic decisions as to the location of ports, capacity, utilization, wharfage, storage facilities, and United States Transportation Command (USTRANSCOM) headquarters as stated in FM 100-10-1. The ASCC Assistant Chief of Staff (ACS), movements, is responsible for operating ports and furnishing liaisons with the Navy, Coast Guard, and other interested military and authorized civilian agencies, both of allied countries and the United States. The ACS, Movements, requests, advises, and makes recommendations concerning the engineer troops employed and the work concerned. All engineer branches use the guidance in annex L, Environmental Considerations (joint OPLAN/OPORD), to achieve operational objectives, while minimizing the impact on health and the environment.

5-4. The CCDR may assign construction support responsibilities to Army, Navy, and/or Marine Corps engineer units, depending on their availability and the overall situation. Mutually supporting or follow-on construction must be coordinated with other engineer units assigned to or projected for the AO. See figure 5-1, page 5-4.

ARMY SERVICE COMPONENT COMMANDER RESPONSIBILITIES

5-5. The functions of the ASCC under 10 USC, for the construction or rehabilitation of a port may include—

- Studies of intelligence reports and all available reconnaissance applying to each port area that is considered for use.
- Tentative determination of the ports or coastal area to be used as a part of overall strategic planning.
- Assignment of the mission of the port.
- Determination of port requirements.
- Tentative decision on the general methods of construction to be used and the determination of engineer units, special equipment, and materials required.

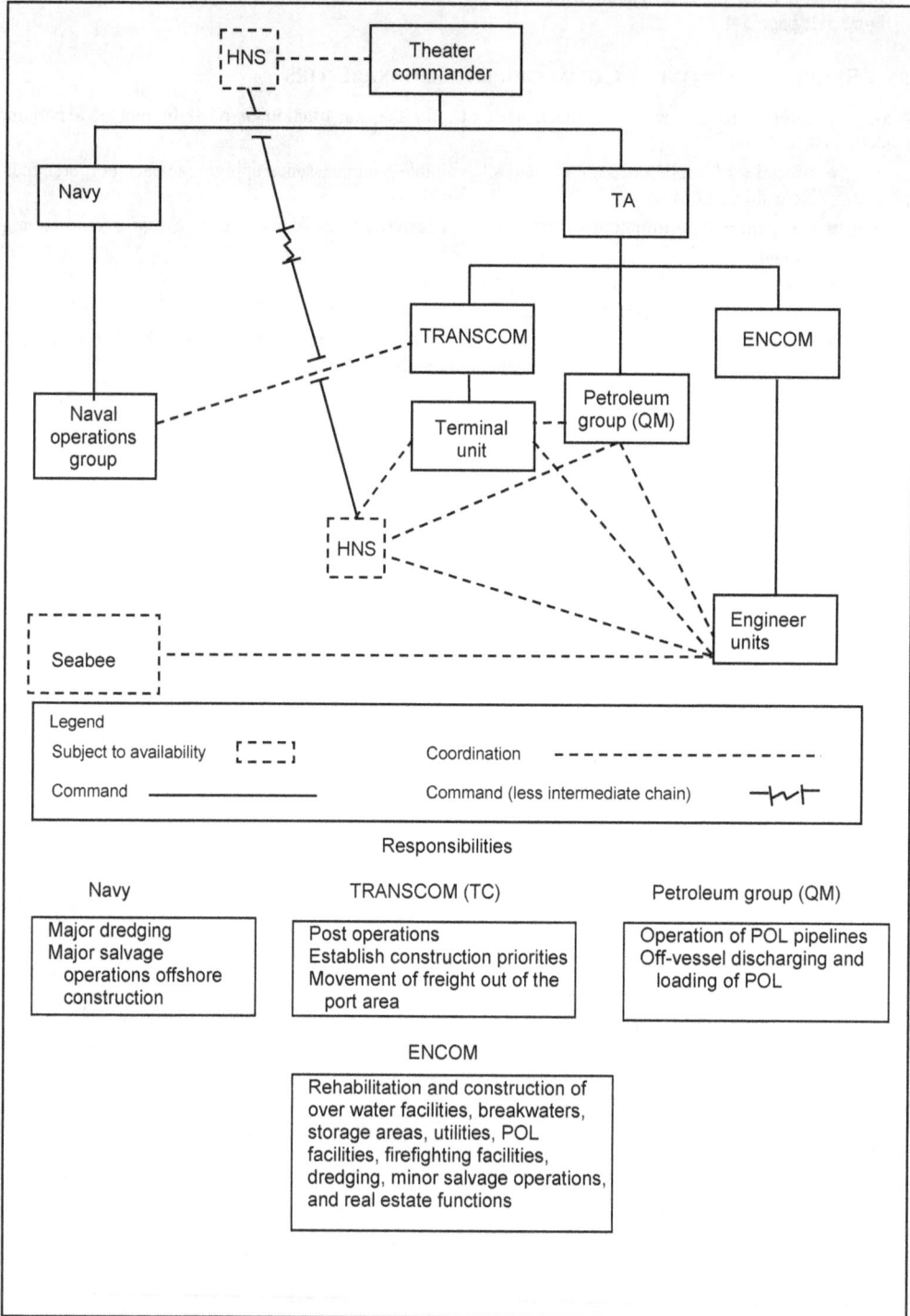

Figure 5-1. Port construction command and coordination

Navy Responsibilities

5-6. The Navy has many of the same capabilities for port construction as the Army. The Navy accomplishes its port construction missions with Seabees. Close coordination between the forces must be done to avoid duplication or counterproductive efforts.

Marine Responsibilities

5-7. The Marines have a considerably smaller overall engineer force than the Army. As such, they have a smaller role, though no less important, during most port construction operations. The majority of Marine engineer forces is primarily task-organized to support maneuver units and may only provide port construction support sufficient to move Marine units through a port.

Army Engineer Units

5-8. Army engineer units are responsible for port construction and rehabilitation and for coordinating all work with that of any Navy units engaged in harbor clearance and salvage operations, such as the neutralization of mines and underwater obstacles. Engineers perform minor salvage operations, such as clearing obstructions and debris from harbor entrances and improving channels. This does not include large-scale salvaging, which is a Navy responsibility. Vertical and horizontal companies augmented with a concrete section, dive team, and other specialty teams and sections accomplish the majority of the tasks. In performing their mission of rehabilitation, construction, and maintenance of a port, Army engineers are responsible for—

- The construction and repair of breakwaters, docks, piers, wharves, quays, moles, and landing stages.
- The construction and maintenance of roads in the port area.
- The construction and major maintenance, only, of railway facilities required by the port.
- The construction of storage and marshaling areas required by the port.
- The construction or reconstruction of port utilities, including water supply, electric power, and sewerage, if required.
- The construction and major maintenance, only, of tanker unloading facilities, including mooring facilities, submerged pipelines, surface pipelines, and rigid petroleum, oils, and lubricants (POL) tank farms.
- The maintenance and operation of the firefighting facilities of the port.
- Any dredging, except as accomplished by the Navy.
- The debris and EH clearance in the port area.
- The acquisition of buildings, facilities, and other property within the port area for military use.
- The provision for warehouses, depots, quarters for port personnel, and other facilities as required for the operation of the port.
- The continuous study of the port situation and the preparation of tentative plans for possible contingencies.
- The requisitioning of supplies and equipment to carry out the mission.
- The provision for diver support.
- The liaison with naval units to coordinate construction with harbor clearance activities.
- The recommendations for real estate allocations.
- The recommendations based on environmental considerations, to include force health protection (FHP).
- The advising of the CCDR and staff on engineering matters connected with the identification, classification, in-transit storage, movement, and distribution of engineer equipment and Class II and Class IV construction materials.

5-9. The engineer unit with overall responsibility for port construction or rehabilitation may be the theater engineer command, engineer brigade, battalion, or company depending on the scope of work. For port construction, it is essential to task-organize a modularized force with the right type and number of companies, platoons, sections, and teams. The modularized force may include horizontal and vertical companies, concrete sections, heavy dive teams, pipeline companies, survey design sections, and other units as the mission requires.

5-10. Tele-engineering technology improves global reach, letting engineers who are deployed reach back via satellite to experts for additional advice and expertise as necessary (see appendix B). This reduces the number of engineer footprints while providing the technical expertise necessary to support GE projects, such as port construction, rehabilitation, and repair.

TRANSPORTATION UNITS

5-11. Transportation units are responsible for operating the port. The unit coordinates operational activities with the completion of necessary projects, and provides liaison with the Navy and Coast Guard. The transportation unit also performs a continuous study of the needs of the port facilities to ensure smooth and orderly flow of personnel, supplies, and materiel through the port. The unit staff plans, supervises, and controls freight movement from the port by rail, motor, and inland water transportation, and under certain conditions, air transport. Finally, the transportation unit is responsible for establishing engineer construction priorities.

QUARTERMASTER UNITS

5-12. Quartermaster (QM) units have responsibility for the operation of petroleum pipeline systems including off-vessel discharging and loading. They coordinate with naval units, engineer units, and transportation units in determining the location of tanker unloading and vessel fueling facilities.

HOST NATION AND LOCAL LABOR SUPPORT

5-13. HNS is used to the fullest extent to reduce the requirements for engineer units and to expedite construction. In the rehabilitation of developed areas, it may be practical to arrange employment of HN engineers, contractors, and superintendents with their organizations. These may include a variety of skilled workers. In many undeveloped AO, local businesses have established organizations to employ and supervise labor in agriculture and other pursuits. Such organizations can often provide labor skilled in primitive construction methods. In either case, the plans for employing civilian labor must include adequate consideration of such factors as housing, transportation, local customs, language difficulties, any locally determined complications due to race or religion, and adapting construction plans to the methods and materials to be used. The use of local civilian labor may result in savings in mobilization and demobilization costs and additional savings due to the local wage scale.

PLANNING FACTORS

5-14. Wherever possible, port construction efforts in the OE are based upon the rehabilitation and/or expansion of existing facilities rather than new construction. Once the decision as to the location of the port has been made at theater headquarters, the mission is assigned to an appropriate engineer level of command. The location of the port will be made based upon an analysis of the projected capacity of the facility, the quantity and nature of cargo to be handled, the tactical and strategic situation, and the construction materials and assets available.

5-15. Careful planning based upon extensive and detailed reconnaissance is essential to successful port construction. This reconnaissance should begin upon receipt of the mission and continue throughout construction and up to actual occupation. A thorough initial reconnaissance will help planners estimate logistics requirements by providing data on the physical condition of the port to be seized or occupied. Geospatial products may assist before, during, and after the reconnaissance has been completed.

5-16. Based upon this analysis, construction assignments, facilities required, and scheduled target dates for various phases of development are derived and outlined in the OPORD. From this information, a construction schedule is formulated. Construction schedules are prepared to show in detail the time plan for all operations in their proper sequence. Equipment hours and man-hours of labor required for each principal operation are then tabulated. The construction schedule is based on the—

- Time allowed for completion.
- Available equipment.
- Type of labor available (regular troop units, reserve troop units, newly activated troop units, local contractors, and international contractors).
- Delivery of construction materials.
- Local sequence of operations.
- Necessary delays between operations.
- Weather.
- Protection and AT considerations and assessment of the threat.
- Environmental and health considerations.

5-17. After the port has been occupied, planners must carefully and critically examine previous plans in view of the actual physical condition of the port. The impact of proposed changes on logistics and scheduling must be coordinated through engineer, transportation, and command channels. Priorities established in the OPORD may have to be modified after construction is undertaken. Planning and scheduling are based on meeting all immediate needs, while ensuring that all work contributes toward the anticipated requirements.

5-18. Studies are made to determine the relative value of rehabilitation and construction. These studies compare the value to be gained from specific facilities within a port to the construction effort required. Among other factors, selection of the best ports for further development is determined by the need for dispersion, location of logistics requirements, time and effort required to move construction units, local availability of materials, and civilian labor.

5-19. The Army theater sustainment command estimates port capacity requirements. The engineer usually makes an independent estimate of the capacity of the port under various alternative methods of construction, repair, or rehabilitation. This procedure serves as an aid to determine the most advantageous relative priorities of engineer projects. The capacity estimates of the sustainment brigade or theater opening element, however, must govern with respect to military loads. On the basis of port capacity estimates, the engineer recommends schedules for construction and rehabilitation of port cranes and other facilities, road and railroad construction within the port area, preparation of storage and marshaling areas, and the like. Some considerations in port capacity estimating and planning are shown in the following paragraphs.

UNIFIED FACILITIES CRITERIA

5-20. The UFC system provides planning, design, construction, operations, and maintenance criteria and applies to all service commands having military construction responsibilities. It will be used for all service projects and work for other customers where appropriate. The UFC handbooks are a guide for engineers, planners, and facility personnel in scheduling, inspecting, maintaining, and repairing mooring hardware at waterfront facilities and related facilities. The following UFC apply to port construction:

- UFC 4-150-02.
- UFC 4-150-06.
- UFC 4-150-07.
- UFC 4-150-08.
- UFC 4-151-10.

WHARF FACILITIES

5-21. Rehabilitation and construction priorities, choice of construction materials, and operational plans for the port are factors which determine the attainment of the greatest capacity from the wharfage with the least expenditure of manpower and material. Port capacity estimates are based on the discharge rates of ships, either at the wharf or in the stream. Priority is given to the methods which allow ships to be discharged more quickly. Construction is scheduled in coordination with transportation operations so that construction activities interfere as little as possible with the discharge of ships.

ANCHORAGE AVAILABLE

5-22. When sheltered anchorage is available, lighterage operations offer a means of discharging cargo while deepwater wharves are under construction or repair. By conducting lighterage operations while construction and rehabilitation work go forward, continued unloading is possible through the use of the following alternatives:

- The continuous dredging of the deepwater wharf approach channel by using a shallow-draft approach and discharge outside of dredging work areas.
- The use of shallow-draft parts of the wharf systems while some of the deepwater wharves are under construction.
- The unloading of shallow-draft vessels over deep-draft wharves during construction.

5-23. Planners may use the basic periods of time, such as the two-shift, 20-hour working day, or the days in a month to prepare estimated labor needs extending over a period of time. However, adverse physical conditions peculiar to the location must be considered. For example, severe icing conditions during the winter months, periods of extreme tide range, or severe seasonal winds may have a direct bearing upon construction or rehabilitation work. When heavy seasonal rains, snowfall, icing, seasonal winds of unusual severity, frequent or seasonal fogs, or exceptionally high or low temperatures are typical to a coastal area, work time estimates should be modified to allow for such conditions.

5-24. Good engineering design is based on careful consideration of pertinent variable relationships and their applications. A temporary or expedient construction design is good if it fulfills its purpose within job limitations. Whenever possible, standard designs are used to save time in design, construction, and maintenance. Standard designs and their accompanying BOM are the basis for advance procurement of construction materials and equipment. The engineer must fit these designs to the site and adapt them to the existing conditions. Reconnaissance, construction surveys, soil bearing tests, driving of test piles, and perhaps a sieve analysis of local sands and gravels are prerequisites to the preparation of final design drawings and BOM. Design of nonstandard structures is usually carried out only if standard designs cannot be adapted.

5-25. FM 101-10-1 gives planning factors for approximate materials and man-hour requirements in overall planning and estimating of general and break-bulk cargo port construction. TM 5-301-2, TM 5-301-4, and TM 5-303, also give data on design, material, and labor requirements for port structures.

PORT CONSTRUCTION

PHASED CONSTRUCTION

5-26. Current procedures for port construction in undeveloped areas usually fall under the following phases:

- **Phase 1, preliminary.** This phase includes all requirements from the arrival of construction units to the beginning of construction of deep-draft wharves. LOTS operations are conducted during this phase.
- **Phase 2, initial construction.** This phase continues to the point at which the first cargo ship berth is fully operational, including road and rail connection, water supply and electrical services, and bulk POL handling facilities that can receive liquid fuels direct from oceangoing tankers.
- **Phase 3, completion.** This phase ends when all authorized facilities are fully operational.

CONSTRUCTION METHODS

5-27. Commercial records indicate that at least 9 months are required for a skilled construction crew of 30 to construct a modern (about 80 by 1,000 feet) steel or concrete pile wharf by conventional (cast-in-place and/or on-site job erection) methods. This time requirement, even allowing for larger construction crews, indicates that neither steel nor concrete pile wharves will likely be built by conventional methods in the future. Recent studies indicate that although steel and concrete will be the most common building materials in new military port construction, their use will probably be limited to new, unconventional construction methods.

STEEL WHARVES OR PIERS

5-28. The use of steel in future military port construction is expected to occur mainly in the construction of expedient container ports with large self-elevating, self-propelled, and spud-type barge pier units. These can be put into service in relatively short periods of time.

5-29. These structures have been used extensively in the oil exploration industry. Their recommended use in expedient port construction is based not only on concepts, but on actual use in situations at least as demanding as those found in modern military operations. The newer versions of these barges use truss-type supports rather than caissons. They may be elevated at a much faster rate (50 feet per hour) and are more relocatable than the older DeLong-type piers (figure 5-2, page 5-10). This capability may limit the planning for construction and expansion of future ports to getting the individual components to the jobsite.

CONCRETE WHARVES OR PIERS

5-30. Commercial port engineers have prepared and are continuing to prepare designs for precast concrete pier pilings, caps, decks, and curbs. These techniques should reduce conventional concrete port construction time requirements considerably.

CONSTRUCTION MATERIALS

5-31. Materials demanded for port construction are often quite specialized or unique. Class IV supplies include all construction materials and installed equipment. Theater requisitions for engineer construction materials must take account of project requirements for special large-scale operations. Issues from stocks are based on the requirements for the particular work on which the requisitioning unit is engaged. Critical items of Class IV supply may be issued under policies approved by the G-4; uncontrolled items are issued on call.

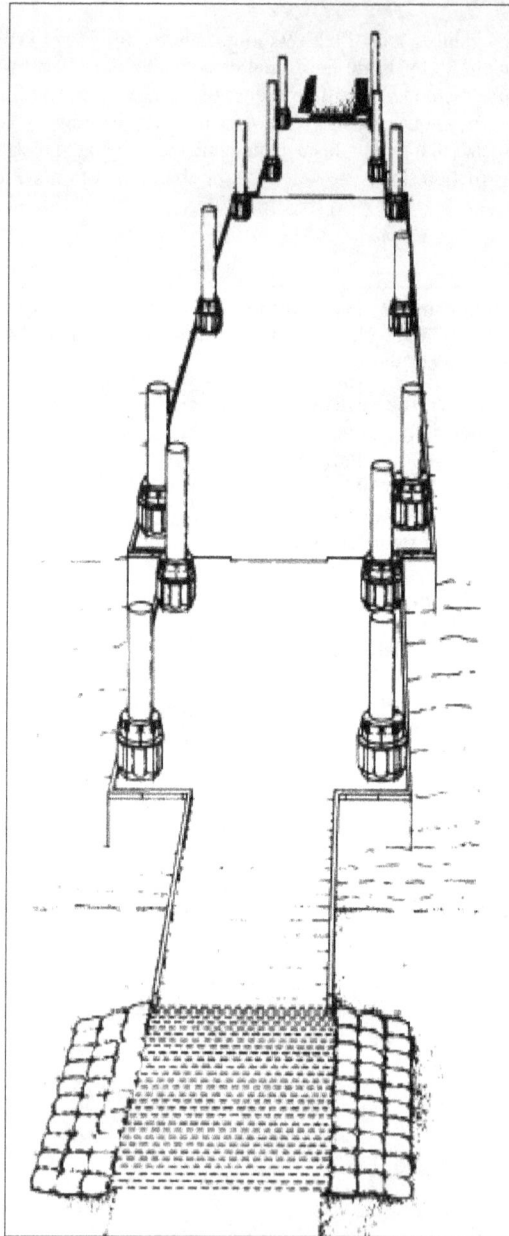

Figure 5-2. DeLong pier

5-32. The task of providing engineer construction supplies is so large, complex, and costly, that every effort must be made to simplify it through the use of local procurement. The unit supply officer locally maintains a continuous inventory of stocks of construction materials and equipment available. Class IV supplies suitable for local procurement may include—

- Lumber.
- Cement.

- Structural steel.
- Sand, gravel, and rock.
- Plumbing supplies
- Electrical supplies.
- Hardware.
- Paint.

SUPPORT FACILITIES

5-33. A large amount of construction effort goes into building port support facilities. If a port is located in an area where there is an adequate rail or roadway network, cargo-handling (break-bulk or container) operations will be more efficient when there are like connectors on the wharves. Engineer units are responsible for the construction of rail and roadway facilities required by the port. Plans are worked out in coordination with the Transportation Corps (TC) requirements.

5-34. Designs currently being recommended to the Army for future expedient military container port construction generally specify tractor-trailers to transport the individual containers from the wharves. The wharf must be of sufficient strength (capable of supporting up to 1,000 pounds per square foot of live loads) and width (usually 80 to 100 feet) to accommodate fully loaded tractor-trailers and must be constructed to an elevation from which suitable connections can be made to existing or planned roadway networks.

5-35. Other on-shore construction requirements include—
- Potable and nonpotable water supply for ships docked or moored in the port and the port itself.
- Electric power supply and distribution, which may require overhead and underground systems.
- Firefighting facilities and special systems as needed, such as special facilities for POL terminals.

5-36. Suitable water depths must be maintained at ports. According to FM 101-10-1, a minimum low tide water depth of about 33 feet should be used for planning purposes because it will accommodate virtually all deep-draft vessels. However, the recent trend toward containerization and the use of large tankers with over 50,000 hundred weight capacities indicate that some future military ports should be planned with minimum water depths of 40 to 50 feet. The planned construction of wharves in water depths several feet less than desired may also be justified where—
- It is established that the required depth can be obtained by dredging, that such dredging is practical as part of the construction project, and that it can be performed without endangering the in-place wharf structure.
- Short-term use is anticipated, making lighterage more feasible than dredging or wharf relocation.
- The actual minimum water depths of new wharf construction are dictated by the wharf's intended use (POL wharf, container wharf, or lighter wharf). These depths are determined and given in the OPORD and/or construction directive.

5-37. Dredging may be required to establish and maintain the required depths. Experience gained during World War II and in Vietnam indicates that there are a number of specific problems associated with dredging projects in an AO. Transportation of dredges to the AO can be difficult. Hopper dredges and side-casting dredges are the only ones that are seagoing. Other dredges must either be towed to the site or assembled from components transported aboard cargo ships.

5-38. It is difficult to secure dredges within the AO. The routine patterns followed by dredges limit the effectiveness of any passive defensive measures. Pipeline dredges are virtually stationary targets. The availability of dredges and crews for use in early stages of deployment in an AO is a major problem. The Army at the present has no trained military dredge crews or portable dredges suitable for use in an AO. USACE has dredges, but the availability of these is not certain and must be planned for and requested well in advance.

5-39. Sweeping, covered in detail in TM 5-235, is a method of locating pinnacles or other obstructions, which exist in navigation areas above the depth limits required by the draft of the largest ships to use the area. Sweeping is always used as a final check after dredging operations.

PORT REPAIR AND MAINTENANCE

5-40. Repair and maintenance involves the correction of critical defects to restore damaged facilities to satisfactory use. Repair and maintenance of conventional and expedient construction could include emergency repair, major repair, rehabilitation of breakwater structures, and expedients.

EMERGENCY REPAIR

5-41. Emergency repair is work done to repair storm, accident, or other damage to prevent additional losses and larger repairs. Emergency repairs include—

- Repairing breached breakwaters to prevent further damage to harbor installations.
- Repairing wharf damage caused by ship or storm damage or enemy action to restore structural strength.
- Dumping rock to control foundation scour or breach erosion.

MAJOR REPAIR

5-42. Major repair is significant replacement work that is unlikely to recur. Major repair includes—

- Replacing wharf decks.
- Resurfacing access roads and earth-filled quays.
- Replacing wharf bracings and anchorages which have been destroyed by decay or erosion.
- Replacing an entire spud barge pier, a spud, or other major barge pier accessories.

REHABILITATION OF BREAKWATER STRUCTURES

5-43. The repair of breakwaters and similar structures is required to protect the characteristics of a harbor. Breached breakwater structures are repaired by dumping rock of sizes suitable for use in mounds.

EXPEDIENTS

5-44. The use of expedient methods should be encouraged during limited port operations while major repair and rehabilitation go forward. A number of possible measures to speed repairs follow:

- Use launches or tugboats with a line to the shore for various hauling and hoisting functions in construction work at the waterfront.
- Erect a derrick or install a crawler or truck-mounted crane on a regular barge; landing craft, mechanized (LCM); a barge of pontoon cubes; or a barge fabricated for military floating bridge units may improvise a floating crane.
- Fabricate rafts for pile-bent bracing operations from oil drums, heavy timbers, spare piles, or local material.
- Improvise floating dry docks for small craft from Navy pontoons.
- Improvise light barges, floating wharf approaches, and small floating wharves from steel oil drums.
- Use diagonal flooring laid over existing decking which strengthens a structure by distributing the load over more stringers.
- Remove the decking for adding stringers, or place smaller stringers on the pile cap between existing stringers from beneath the decking and wedged tight against the deck.
- Take up several floor planks and drive piles through the hole if the wharf can support the weight of the pile driver. Cap new pile bents and wedge them tightly against the stringers.

- Use a rock or ballast-filled timber crib to replace a gap in a pile wharf structure or to extend the offshore end on the wharf. The timber crib may be built on land, launched by using log rollers, floated into position, and filled with rock or ballast to hold it in place.
- Use standard military floating bridges or Navy pontoons to supplement or temporarily replace damaged causeways.
- Use standard military floating bridges or Navy pontoons to provide access between undamaged sections of off-loading piers.
- Ensure that when a section of a wharf has been destroyed, the face of the wharf is restored first so that ships may be worked while the area behind the face is being restored.
- Use the shore end of a pier for lighterage or other short vessels while the pier is being extended.
- Bridge part of a solid-fill wharf using standard or nonstandard fixed bridging (see FM 3-34.343).
- Fill a slip with rubble so that ships cannot be brought to the face of the wharf.

Note. It may be possible to fend ships off with camels, barges, or other devices so that they will be retained in deepwater for unloading. Alternatively, it may be possible to use standard trestles, fixed bridging, and assembled Navy pontoons to extend the width of the pier.

- Use the hull of a capsized or sunken vessel as the substructure for a pier.
- Anchor the shore end of a causeway constructed from Navy pontoon cubes onshore by excavating a section of beach, floating the pontoons into the temporary inlet thus made, and then backfilling to provide a solid anchorage.

LOGISTICS OVER-THE-SHORE OPERATIONS

5-45. At least 90 percent of the tonnage required to support deployed forces in the AO must be provided by sea LOC. Although air LOCs will usually carry high priority shipments and personnel, sea LOCs will likely bear the main burden. The uninterrupted delivery of materiel requires that vulnerable fixed port facilities be backed up by a flexible system, and LOTS operations provide that system. Armed forces LOTS operations involve transferring, marshaling, and dispersing materiel from a marine to a land transport system. The rule of thumb for planners is that 40 percent of all cargo entering contingency theaters by surface means will need to be delivered through LOTS terminals. In some theaters, this proportion may be much greater. Beaches distant from fixed port facilities serve as LOTS sites. The rapid establishment of a viable LOTS system depends on engineer construction and maintenance support.

RESPONSIBILITIES

5-46. Logistics planning to support deployed forces on a foreign shore always begins with an evaluation of in-place fixed port facilities and capacities. These, combined with connecting railway, highway, and inland waterway networks, are the major logistic systems required for military operations. When a reckoning of available resources is complete, planners determine the need for LOTS terminals to supplement and back up the transportation net.

5-47. Overall responsibility for LOTS operations lies with the TC. Each LOTS terminal acts under the direct control of a transportation terminal battalion made up of two service companies and appropriate lighterage units. The CCDR may assign construction support responsibilities to Army, Navy, and/or Marine Corps engineer units, depending on their availability and the overall situation. Mutually supporting or follow-on construction must be coordinated with other engineer units assigned to or projected for the AO. Army engineers must be prepared to support the LOTS mission because—

- Existing ports may be damaged, incomplete, or unavailable.
- Existing ports may be unable to handle resupply operations.

- Existing port facilities are vulnerable to enemy activities, such as mining, CBRN, and air interdiction.
- Existing ports under repair may be unavailable for long periods.

5-48. Engineer units give construction, repair, and maintenance support to LOTS operations. An engineer unit may expect to encounter these missions in supporting a LOTS operation by—

- Constructing semipermanent piers and causeways.
- Preparing and stabilizing beaches.
- Constructing access and egress routes from beaches to backwater areas.
- Constructing access to marshaling areas and/or adjoining LOTS sites.
- Constructing marshaling and storage areas.
- Constructing road and rail links to existing LOCs.
- Constructing utility systems.
- Constructing POL storage and distribution systems.
- Providing other assistance or maintenance determined by the terminal commander.

LOGISTICS OVER-THE-SHORE INSTALLATION

5-49. Initial LOTS planning and site selection are coordinated between the theater opening element or sustainment brigade commander (TC) and the Navy or Military Sealift Command (MSC). Initial selection is based on map studies, hydrographic charts, and aerial reconnaissance.

RECONNAISSANCE

5-50. The reconnaissance party includes representatives of the terminal group commander, the terminal battalion command, the supporting engineer, the supporting signal officer, military police, and Navy personnel to provide advice on mooring areas. Others participate if the situation dictates, or at the terminal commander's request. The reconnaissance party briefs the terminal commander on its findings. The briefing must cover—

- Engineer efforts required when preparing and maintaining the site, based on available units, equipment, and materials.
- Signal construction and maintenance required for necessary communications within the beach area, and between the beach and the terminal group headquarters.
- Lighterage craft (landing craft utility [LCU], LCM, and light air cushion vehicle-30 [LACV-30]) types that may be used based on beach conditions.
- Safe havens for lighterage craft in stormy weather.
- Location and desirability of mooring areas.
- Adequate egress from the beach. Availability of the egress and the beach dimensions are key factors in determining the tonnage capacity for the beach.
- Tidal range.

5-51. See Figure 5-3. The supporting engineer must be informed about the layout of the LOTS site, because the layout determines the required engineer effort. A LOTS layout varies with the situation and existing geographic conditions. The physical size of the individual site depends on the security considerations, soil trafficability, the number of ships to be unloaded at the site, and the type of cargo coming ashore. For example, a LOTS terminal may need to be very large if ammunition and/or POL are being unloaded over a beach that is subject to enemy attack. General cargo unloaded over a secure beach needs less area.

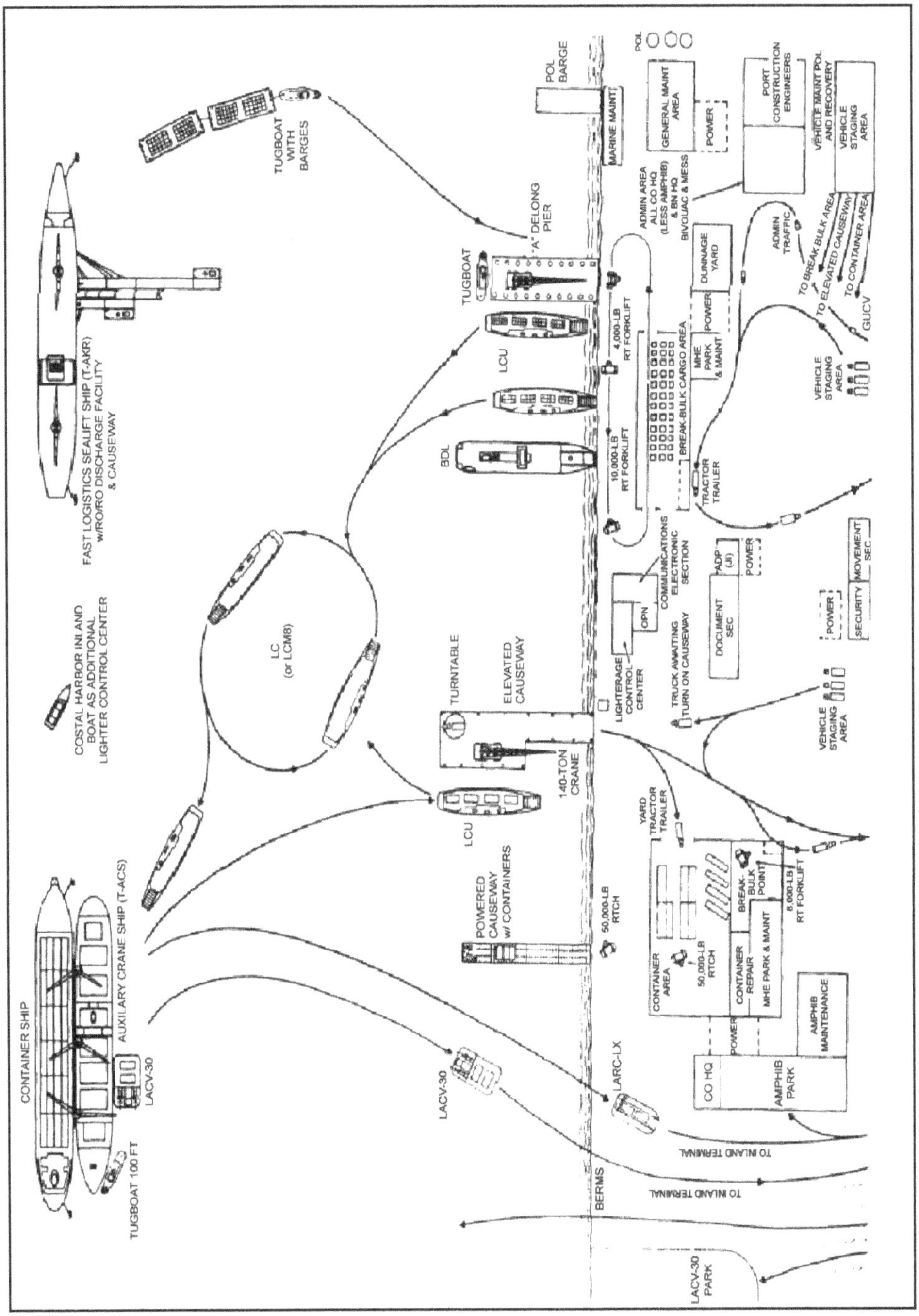

Figure 5-3. Typical LOTS operations

LANDING CRAFT UNLOADING POINTS

5-52. Knowledge of the beaching positions designated for landing craft is important to the supporting engineer, especially if landing points are to be used for extended periods. A common maintenance problem on beaching positions is the creation of troughs or pits in the beach beyond the waterline. Troughing is caused by landing craft ramps, which dig into the inclined beach at a steep angle. This problem is exacerbated when wheeled vehicles dig into the sandy beach material, and water washes the loosened material away. Vehicles can easily bog down and stall in these troughs, slowing unloading operations. Engineers can reduce the troughing by placing stone or gravel at the unloading point, or by cutting down the slope of the beach. Both these measures require maintenance for as long as the unloading points are used.

PIER AND CAUSEWAY CONSTRUCTION

5-53. FM 5-134 provides data on the use of pile material in the construction of piers and causeways. Piers and causeways allow cargo vessels direct beach access, eliminating multiple handling of material and speeding unloading times. Piers are structures with working surfaces raised above the water on piles. Piers project beyond the surf zone. Their stability and protected working surfaces permit unloading at times when wave action would otherwise prevent landing craft from operating across a beach. The DeLong pier is a self-contained pier that can be brought to the LOTS site and emplaced in a relatively short time. Specially trained engineer personnel from the port construction company and certain other units can install this equipment. Other engineer assistance is required at the beach end of the pier to prepare the beach and anchor the pier. Causeways are floating structures, which project out from the beach. In some applications, they are used as rafts to ferry equipment from ship to shore. Causeways are more susceptible to wave action than piers, but they are more easily deployed. In areas where wave action is not a significant problem, causeways can be used as floating piers. Engineers provide beach preparation and anchoring for causeway operations.

ROAD CONSTRUCTION

5-54. The major engineer effort in LOTS is invested in road construction and maintenance. Considerable effort must be spent to stabilize soil and improve trafficability in the beach area. Constructed roads must withstand the impact of materials handling equipment (MHE) carrying extremely heavy loads. Roads that support LOTS are usually constructed in a loop to reduce their required width and eliminate vehicle turning as much as possible (see chapter 7).

5-55. The availability of construction material determines the types of roads that can be constructed. Naturally occurring materials, such as rock and wood, may be scarce or of poor quality. Portland cement may not be available or may be prohibitively expensive to use. Sand grid material is excellent for use in areas of cohesionless soil. Matting and/or steel planking may be used if they are available (see FM 55-60 and figure 5-4 for more details). When roads are constructed in areas of poor soil conditions, roadways must be well marked and adequate drainage must be provided.

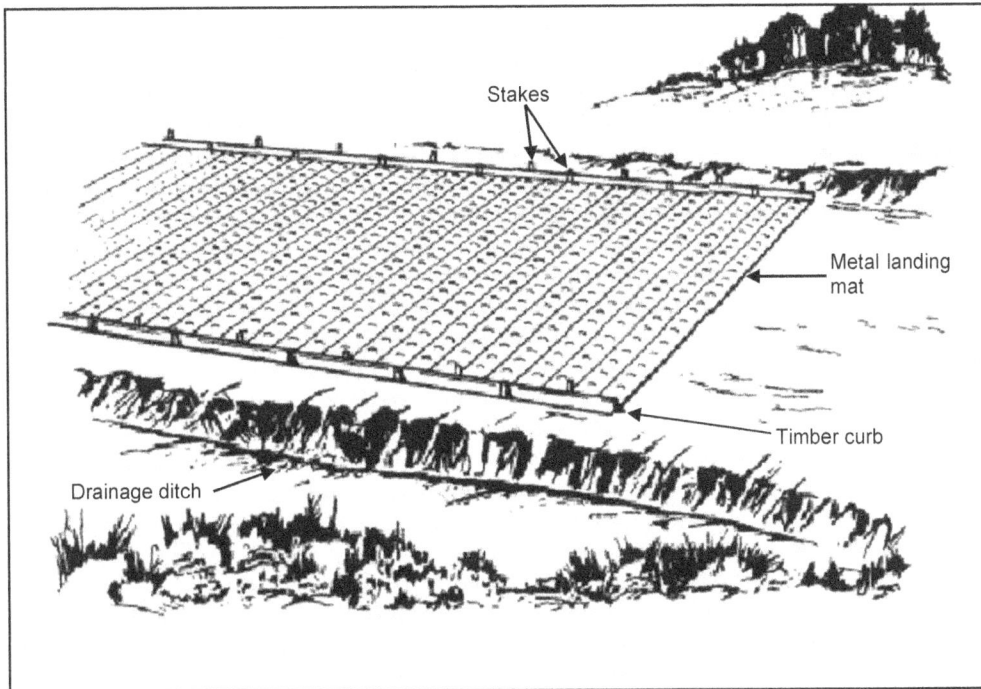

Figure 5-4. Field expedient matting

MARSHALING AREAS

5-56. Marshaling areas (figure 5-5, page 5-18) serve as a collection point from which unloaded materials and equipment can be distributed to the proper units. The size of the marshaling area varies with the size and type of shipping, the unloading rate, the hostile situation, and the units being supported. Marshaling areas vary from 25 to 500 acres. In hostile environments, marshaling areas are dispersed with acreage divided into many small parcels. Other protection considerations must be integrated into the design of marshaling areas as well.

5-57. Marshaling area surfaces must be stable enough to support a loaded piece of MHE. MHE with loads may weigh up to 100,000 pounds. Access and egress roads must be capable of supporting the same loads. The surface must be protected with adequate drainage.

5-58. Most material shipped to an OE via surface transportation is containerized. Once ashore, the containers are opened and unpacked for distribution to the intended units. Empty containers are collected and reloaded aboard ships, then returned to their point of origin. Container collection areas must be provided. These areas must have the same trafficability, drainage, and access and egress characteristics as the marshaling areas, and can require nearly as much space.

AMMUNITION STORAGE AREAS

5-59. Areas where ammunition is to be unloaded, sorted, and temporarily stored requires the same type of planning and engineer effort as marshaling areas. In addition, engineer units will have to perform more horizontal construction work, making earthen berms and revetments. Ammunition supply points (ASPs) that will be used for an extended time must be provided with overhead protection from the elements.

5-60. Ammunition storage areas must be remote from other activities on the beach. They must be dispersed and camouflaged. Each site requires access and egress routes, preferably arranged so that vehicles will not have to back up.

PETROLEUM, OIL, AND LUBRICANTS STORAGE AREAS

5-61. Fuel storage areas on the beach will likely be the largest concentration of fuels in the distribution system. Construction of rigid storage tanks and distribution pipelines within the storage areas is a major engineer task. Engineers also support the QM Corps by installing collapsible tank farms and related facilities.

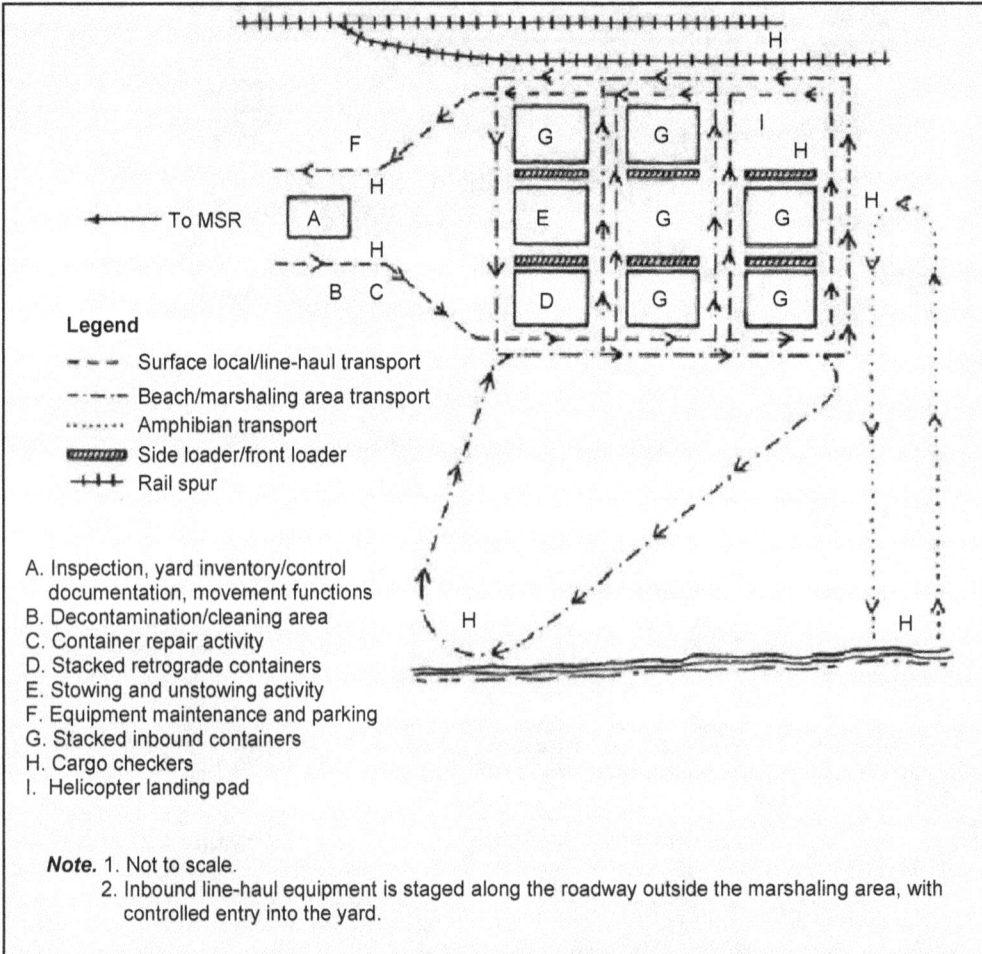

Figure 5-5. Container yard marshaling area

Chapter 6
Airfields and Heliports

Air power is a thunderbolt launched from an eggshell invisibly tethered to a base.
Hoffman Nickerson, Arms and Policy (1945)

Airfields and heliports support the deployment, operational maneuver, and sustainment of forces. An adequate aerial LOC network is one of the keys to shaping operations in today's OE. Airfields and heliports are built, upgraded, repaired, and maintained to meet mission and operational requirements. Depending on mission, enemy, terrain and weather, troops and support available, time available, and civil considerations (METT-TC) aerial LOC may form the primary means of sustaining a contingency operation. Contingency operation airfield and heliport planning involves much more than just the airfield layout (geometry) and pavement structure. It involves planning for all of the supporting facilities and infrastructure, such as the air traffic control and landing system (ATCALS), and POL and munitions storage facilities needed to sustain airfield operations. It also involves protection, security measures, health, safety, and environmental factors. Forward aviation combat engineering prepares or repairs expedient landing zones (LZs), forward arming and refueling points (FARPs), landing strips, or other aviation support sites in the forward combat area and are considered combat engineering tasks and are focused on providing support to tactical combat maneuver forces. See Air Force Engineering Technical Letters (ETLs) 04-7 and 97-9; FM 3-34.2; and FM 5-430-00-2, Volume II. All other airfield and heliport construction are considered GE tasks. Contingency operations airfield planning is similar to base camp or force bed-down planning in many respects. This chapter will focus on just the airfield and heliport aspects (see chapter 11 for details on base camp and force bed-down).

Note. The Marine Corps and joint doctrine use METT-T without "civil considerations" being added.

RESPONSIBILITIES

6-1. Army, Air Force, Navy, and Marine engineers all have the capability to design, plan, construct, upgrade, repair, and maintain airfields and heliports. Army engineers are often responsible for initial airfield damage repair (ADR) as a part of a forcible entry operation. This is a type of the forward aviation combat engineering that is performed by combat engineers to enhance mobility. Army engineers may assist other engineers as directed in airfield and heliport design, planning, construction, repair, and maintenance. The Army provides the following construction support to Air Force-controlled airfields:

- Develops engineering design criteria, standard plans, and material to meet Air Force requirements.
- Performs reconnaissance, survey, design, construction, or improvement of airfields, roads, utilities, and structures.
- Repairs Air Force bases and facilities beyond the immediate emergency recovery requirements of the Air Force (semipermanent and permanent repair).

- Supplies construction materials and equipment.
- Assists in emergency repair of war-damaged air bases.
- Assists in providing expedient facilities (force bed-down).
- Manages war-damaged repair and base development; supervises Army personnel. The Air Force base commander sets the priorities.
- Performs emergency and permanent repair of war damage to forward tactical airlift support facilities.

6-2. The branch of service that is the primary user of the airfield or heliport has the responsibility for certifying that facility for flight operations. In most cases during airfield contingency operations, this is an Air Force responsibility. Air Force engineers may assist other Army engineers, Navy Seabees, or Marine engineers as directed in airfield and heliport design, planning, construction, repair, and maintenance. The Air Force provides the following engineer support:

- Performs primary emergency repair of war damage to air bases and other ADR tasks.
- Constructs expedient facilities for Air Force units and weapon systems. This excludes responsibility for Army base development.
- Operates and maintains Air Force facilities. Air Force engineer units perform maintenance tasks.
- Provides crash rescue and fire suppression.
- Provides hazardous material (HAZMAT) response.
- Manages emergency repair of war damage and force bed-down construction.
- Provides infrastructure support for solid waste and hazardous waste disposal.
- Supplies material and equipment for its own engineering mission.
- Provides the EBS and EHSA for the airfield and its support facilities.

PLANNING

6-3. Airfields and heliports are classified by their degree of permanence and type of aircraft they are designed to support. They are essential for controlling aircraft, either fixed-wing and/or rotary-wing. These controlling aircraft, or aircraft combination, are identified for each kind of facility to establish limiting airfield and/or heliport geometric and surface-strength requirements. For information on survivability (hardening) support, to include the construction of revetments for helicopters, see FM 5-103. For information on Air Force aircraft survivability, see Air Force Manual (AFM) 91-201, Category Code 141-182, Hardened Aircraft Shelters.

6-4. Army airfields and heliports are divided into the following six classes (see UFC 3-260-02, chapter 2):

- **Class I, helipads-heliports with aircraft 25,000 pounds (11,340 kilograms) or less.** The controlling aircraft is a UH-60 aircraft at a 16,300-pound (7,395 kilograms) operational weight.
- **Class II, helipads-heliports with aircraft over 25,000 pounds (11,340 kilograms).** The controlling aircraft is a CH-47 aircraft at a 50,000-pound (22,680 kilograms) operational weight.
- **Class III, airfields with Class A runways.** The controlling aircraft combination is a C-23 aircraft at a 24,600-pound (11,200 kilograms) operational weight and a CH-47 aircraft at a 50,000-pound (22,680 kilograms) operational weight. Class A runways are primarily intended for small aircraft, such as C-12s and C-23s.
- **Class IV, airfields with Class B runways.** The controlling aircraft is a C-130 aircraft at a 155,000-pound (70,310 kilograms) operational weight or a C-17 aircraft at a 580,000-pound (263,100 kilograms) operational weight. Class B runways are primarily intended for high performance and large, heavy aircraft such as C-130s, C-17s, and C-141s.
- **Class V, contingency operations heliport or helipads supporting Army assault training missions.** The controlling aircraft is a CH-47 aircraft at a 50,000-pound (22,680 kilograms) operational weight.

- **Class VI, assault LZs for contingency operations airfields supporting Army training missions that have semiprepared or paved surfaces (also known as forward landing strips).** The controlling aircraft is a C-130 aircraft at a 155,000-pound (70,310 kilograms) operational weight or a C-17 aircraft at a 580,000-pound (263,100 kilograms) operational weight.

6-5. Air Force airfields are classified into six mission categories. A controlling aircraft or combination of controlling aircraft has been designated for each category to establish limiting airfield, geometric, and surface strength requirements. The six airfield categories include (see UFC 3-260-02, chapter 2 and chapter 3) the following:

- Light (F-15 and C-17).
- Medium (F-15, C-17, and B-52).
- Heavy (F-15, C-5, and B-52.)
- Modified heavy (F-15, C-17, and B-1).
- Auxiliary (F-15).
- Assault LZ (C-130 and C-17).

6-6. A bare base airfield is a site with a usable runway, taxiway, and parking area and a source of water that can be made potable. It must be capable of supporting assigned aircraft and providing other mission-essential resources, such as logistical support, services, and infrastructure (supplies, equipment, people, and facilities). This concept requires modular facilities, mobile facilities, utilities, and support equipment packages that can be rapidly deployed and installed. A bare base airfield forms the baseline for contingency operations airfield planning.

6-7. On normal operational airfields, pavements are grouped into the following four traffic areas based on their intended use and design load (see UFC 3-260-02, chapter 2 and chapter 3):

- **Type A.** Those traffic areas that receive concentrated traffic and the full design weight of the aircraft. These traffic areas require a greater pavement thickness than other areas on the airfield and include all airfield runways and, in most cases, runway ends and primary taxiways. All airfield pavement structures on contingency operations airfields are considered Type A traffic areas.
- **Type B.** Those traffic areas that receive a more even traffic flow and the full design weight of the aircraft. These traffic areas include parking aprons, pads, and hardstands.
- **Type C.** Those traffic areas with low-volume traffic, or the applied weight of the operating aircraft is generally less than the design weight. These traffic areas include secondary taxiways, washrack pavements, access aprons, interior portions of runways, and hangar floor areas trafficked by aircraft.
- **Type D.** Those traffic areas with extremely low-volume traffic and/or the applied weight of the operating aircraft is considerably lower than the design weight.

6-8. See FM 5-430-00-2. Based on an airfield's location in the AO, it could be described as a—

- **Close battle area.** Forward airfields are intended to provide focused logistics support and/or support combat missions of short-range , such as attack helicopters and unmanned aircraft systems (UASs) during contingency operations. These airfields are designed to initial or temporary contingency operations standards, depending on mission and operational requirements, and may be paved or semiprepared. These may be initially prepared or repaired as forward aviation combat engineering tasks.
- **Support area.** Intermediate airfields are intended to provide general logistics support, support combat missions of longer-range aircraft during contingency operations, and/or training. These airfields are designed to temporary or semipermanent standards depending on mission and operational requirements. Normally these airfields are paved. These airfields provide a link between close battle area airfields with rear area airfields.
- **Rear area.** Airfields are intended to provide logistics support forward from fixed, secure bases, and support combat operations of long-range aircraft. These airfields are designed to be semipermanent or permanent facilities.

6-9. Most planning factors described in chapter 7 for road designs are applicable for airfields. The most important factors include—

- **Mission.** To achieve a proper design for the airfield, it is essential that the engineer planner have a complete understanding of the number and type of aircraft, purpose, scope, and estimated number of the particular air missions to be flown by the design-controlling aircraft. Categories of missions that may be conducted include reconnaissance, cargo transport, or attack.

- **Enemy.** The engineer planner devises an adequate plan to ensure that construction troops can protect themselves, their equipment, and their materials against harassment and sabotage during airfield or heliport construction. Requirements for additional security forces should be evaluated. The engineer planner should also consider AT and other protection requirements of the aircraft using the airfield and the infrastructure supporting airfield operations. See chapter 9 for more details on protection. See FM 5-103 for those aspects of protection that have to do with *survivability operations* (the development and construction of protective positions, such as earth berms, dug-in positions, overhead protection, and countersurveillance means, to reduce the effectiveness of enemy weapon systems. [FM 5-103]).

- **Terrain and weather.** The engineer planner's attention must be directed first toward selecting sites. Within site requirements dictated by mission and operational requirements, the planner establishes reasonable site requirements for each type of airfield. The planner chooses geographic locations on the basis of topographic features (grading, drainage, and hydrology), soil, vegetation, utilities, climatic conditions, and accessibility of materials. Other site characteristics to be studied include weather patterns (such as temperature, barometric pressure, seasonal variations, and wind directions) and flight path obstacles. The planner evaluates all existing transport facilities to determine the best methods and routes to provide logistics support the project. These include ports, rail lines, road nets, and other nearby airfields that might be used for assembling and moving construction equipment and materials to the construction site. See appendix D for a discussion of the effect of various environmental considerations.

- **Troops and support available.** The planner evaluates the availability and type of engineer construction forces to determine if construction capability is sufficient to carry out the required airfield construction. The planner must weigh the type and availability of local construction materials against the overall needs for proposed construction. Both naturally occurring materials and other possible sources for materials for subgrade strengthening should be examined. Requirements for importing special materials for surfacing, drainage, and dust control must be feasible for available construction time and resources. The planner must have knowledge of the forces dedicated to the ADR. Depending upon base locations, local agreements, and the overall military situation, any combination of Army, Air Force, HN, or contract engineer support may be possible. Time-phased force and deployment data (TPFDD) or population flow into the airfield must be considered when developing the airfield master plan.

- **Time available.** Operational and mission requirements will dictate when the airfields need to be able to support aircraft operations.

- **Civil considerations.** The engineer planner must consider what civilian construction resources are available in the local area and what structures already exist that could be used to support airfield construction, repair, and maintenance. The planner must also consider the environmental impact, restricted areas, political and cultural factors, and other factors that may affect airfield layout and construction.

- **Maximum on ground.** MOG is the maximum number of aircraft that can be accommodated on an airfield. There are two types of MOG. Parking MOG is the total number of aircraft that can be parked at an airfield. Parking MOG is affected by both the overall size of the airfield and by how available space is managed. Working MOG refers to how many or how quickly parked aircraft can be off-loaded, material through-putted from the aerial port of debarkation (APOD), and aircraft serviced and prepared for departure. MHE, trucks, buses, and other surface transport vehicles, road networks, aircraft support equipment, fuel tankers, personnel, and other factors affect working MOG.

6-10. Ideally, working MOG equals parking MOG; when it does not, backlogs occur. MOG is normally expressed in terms of C-141s. A minimum of MOG two is desired for contingency operations airfields. Refer to Air Force Pamphlet (AFPAM) 10-1403 for aircraft dimensions.

6-11. Engineers are responsible for site reconnaissance and recommendations, design of the airfield or heliport, and the actual construction of the individual airfield. Standard designs for the type and capacity of the airfield are available in TCMS. However, the planner must frequently alter these designs to meet time and material limitations or the limitations imposed by local topography, area, or obstruction characteristics. Engineers may alter designs, but must obtain approval for major changes from the user before work starts. The engineer will need to solve the following engineering problems when carrying out most airfield assignments:

- Design a drainage system structure.
- Design runways, taxiways, and hardstands.
- Select or dispose of soils encountered in cuts, determining their usefulness for improving subgrade.
- Choose a method or methods for stabilizing the subgrade.
- Decide upon the type and thickness of the base course.
- Decide upon the type and thickness of the surface course.
- Select the grade for a minimum of earthwork within specification limits.
- Design related facilities, including access and service roads, ammunition and POL storage areas, navigation aids (NAVAIDs), maintenance aprons, warm-up aprons, corrosion control facilities, control towers, airfield lighting, and other facilities.
- Integrate environmental considerations, to include applicable FHP intelligence.

6-12. For airfield planning and design, refer to the following manuals:

- FM 3-34.2 and those associated with forward aviation combat engineering operations.
- FM 5-430-00-1, Volume I.
- FM 5-430-00-2, Volume II.
- TM 5-820-1.
- UFC 3-260-01.
- UFC 3-260-02.
- UFC 4-141-10N.

CONSTRUCTION

6-13. A completed air base is a complex construction project. However, careful planning and a strict focus on essentials can result in a facility that will support air operations soon after construction begins. Subsequent improvements can be made during use. If construction is guided by a master plan, staged completion of each structure can be designed to serve both expedient operation and the final design of the facility.

6-14. Preplanned design layouts within TCMS for each type of field are based on the assumption that previously unoccupied sites will be chosen. However, the layouts have been coordinated so that, within terrain limitations, it is practicable to develop a larger field from a smaller one with minimal construction effort. Existing airfields or a bare base site can be used if they meet minimum requirements or can be upgraded to meet operational or mission requirements. (A *bare base* is a base having minimum essential facilities to house, sustain, and support operations to include, if required, a stabilized runway, taxiways, and aircraft parking areas. A bare base must have a source of water that can be made potable. Other requirements to operate under bare base conditions form a necessary part of the force package deployed to the bare base.)

6-15. It is best to complete an air base to its ultimate design in a single construction program. Often, however, it is necessary to initially design to a lower construction standard to get the base into operation within available time and construction support. In such cases, every effort must be made to proceed to the ultimate design standard for the airfield. Repeated modification of a facility plan is to be avoided.

6-16. A fully completed airfield includes the following types of facilities:

- Runways, taxiways, hardstands, aprons, and other pavements, shoulders, overrun, approach zones, NAVAIDs, airfield marking, and lighting.
- Sanitary facilities (kitchens, dining areas, showers, and latrines).
- Direct operational support facilities (ammunition, storage and distribution of aviation fuels and lubricants, HAZMAT, and waste storage sites).
- Maintenance, operations, and supply (aircraft maintenance, base shops, operations buildings, base communications, photograph laboratories, fire stations, weather facilities, general storage, and medical facilities).
- Indirect operational support facilities (roads and exterior utilities, such as water supply and electric power).
- Administration (headquarters, personnel services, recreation, and welfare facilities).
- General housing and troop quarters.

6-17. The first goal when building an airfield is to achieve operational status. Therefore, construction is designed to support air traffic as soon as possible. The order for construction proceeds according to the following priorities of work:

- **Priority 1, AT and security.** Provide the facilities most essential for air operations as soon as possible. Build airfield operational facilities, such as runways, taxiways, approaches, and aircraft parking areas of minimum dimensions. Provide minimum storage for bombs, ammunition, and aviation fuel. Provide essential airfield lighting, fire protection services, medical, attack warning system, sanitary, power generation, and water facilities. Site in facility groups and ATCALS.
- **Priority 2, improve AT and increase the capacity, safety, and efficiency of all air base operations.** Provide indirect support to operational facilities. Construct access and service roads and essential operational, maintenance, and supply buildings.
- **Priority 3, improve AT and operational facilities.** Provide facilities for administration and special housing such as leach fields, wash racks, landfills, and an explosive ordnance disposal (EOD) range.
- **Priority 4, improve AT and provide general housing.** Institute a base operation and facilities maintenance plan. Sustain environmental and medical surveillance of the airfield and its supporting facilities.

6-18. Construction stages establish a sequence for constructing an airfield. These stages provide for building the airfield in parts, so that minimum operational facilities may be constructed in the shortest possible time. For example, a priority task may be reduced to stages as follows:

- **Stage I.** Stage I provides a loop that permits landing, take off, and circulation and limited apron parking is built. Runway lengths and widths are the minimum required for critical aircraft.
- **Stage II.** Stage II provides a new runway. The stage I runway now becomes a taxiway, and aprons, hardstands, and additional taxiways are built.
- **Stage III.** Stage III provides facilities that are further expanded, and accommodation for more aircraft is added, if necessary. When an existing surface in the rear area is not adequate for all-weather operations in support of heavy transport aircraft or high-performance fighter aircraft, an appropriate pavement structure is designed and constructed.

6-19. Airfield reconnaissance differs from road location reconnaissance in that more comprehensive information is typically required. An airfield project involves more man-hours, machine-hours, and material than most road projects. Air traffic also imposes stricter requirements on traffic facilities than does vehicular traffic. Consequently, the site selected has to be the best available.

6-20. When new construction is undertaken, the planner and the reconnaissance team must choose a site with soil characteristics that meet strength and stability requirements, or a site that requires minimum construction effort to attain those standards. Airfields present more drainage problems than roads. Their wide, paved areas demand that water be diverted completely around the field, or that long drainage structures be built. Sites at the low points of valleys or other depressed areas should be avoided because they tend to be focal points for water collection. As in road construction, subsurface water should be avoided. A desirable airfield site lies across a long, gentle slope, because it is relatively easy to divert water around the finished installation.

6-21. To accommodate missions efficiently, airfields require large areas of relatively flat land. Advance location and layout planning will avoid cramping facilities. To obtain the required area, the airfield may have to be spread over a large area. This may call for a complex network of taxiways and service roads. Runways should be aligned in the direction of the prevailing wind.

6-22. The safe operation of fixed- or rotary-wing aircraft requires that all obstacles above elevations specified by design criteria be removed. This criteria varies according to the operating characteristics of the aircraft that use the airfield. For example, most heliports require an approach zone with a 10:1 glide angle, whereas heavy cargo aircraft in the rear area require a glide angle as flat as 50:1. To achieve the right glide angle, it is often necessary to remove hills and do major earthwork on distant approaches to the airfield. The reconnaissance team should avoid locations that need extensive earthwork to achieve the necessary glide angle. Clearances are also required along the sides of runways. An area of specified width must be cleared of all obstacles and graded according to specification.

6-23. Except for staking requirements, the techniques and principles for conducting airfield and heliport construction surveys are identical to those for roads. An accurate estimate of earthwork volume is essential to proper control and management of a horizontal construction project. Following mass diagram construction and analysis, equipment is scheduled and project durations are determined. Analysis of the mass diagram will also determine haul routes, location of equipment work zones, and areas for waste and borrow sites. Earthwork is conducted as described for road construction in chapter 7, except that project width permits more balancing perpendicular to the airfield's centerline. Earthwork balancing may also occur between adjacent projects (for example, runway and taxiway).

6-24. During construction, permanent drainage structures are essential to the successful completion of an airfield or heliport. Planning considerations are similar to those used for road construction.

6-25. The decision to pave an airfield or heliport during contingency operations is based upon the urgency that the airfield be completed, the tactical situation, the amount and type of anticipated traffic, the soil-bearing characteristics, the climate, and the availability of new materials and equipment. Surfacing must meet the allowable roughness criteria for each type of aircraft that will use the facility. Soil stabilization operations improve strength, control dust, and render surfaces waterproof. The process is discussed in chapter 7.

6-26. Maximum use must be made of existing facilities. However, airfields and heliports may need extensive new support facility construction. Expansion and rehabilitation of existing infrastructure should usually be considered over new construction, since there is generally a substantial savings in time, effort, and materials to upgrade rather than building from scratch. Except in highly developed areas, existing airfields are seldom adequate to handle modern, high-performance aircraft.

6-27. Existing airfield dimensions and pavement structures must be evaluated by the reconnaissance team based on mission and operational requirements to determine if the airfield is capable of supporting air traffic and if not, what construction effort will be required to enable the airfield to meet those requirements. Some airfields may be made adequate with only minimal effort. They may also serve as the nucleus for larger fields that meet the specifications of high-performance aircraft. Helicopters and light planes can often operate from existing roads, pastures, and athletic fields. Combat engineers may be able to upgrade these enough for initial or temporary use through forward aviation combat engineering. Support facilities are converted to standards dictated by the theater construction policy. Imaginative use of existing facilities is preferable to new construction. Ground reconnaissance of an airfield previously occupied by enemy forces must be cautious, since facilities may contain EH. Facility use must be coordinated with HN authorities because existing airfields, particularly in the rear area, are needed by HN air forces and for commercial purposes.

6-28. Priorities for expanding and rehabilitating an existing airfield generally parallel those for new airfield or heliport construction. Procedures, personnel, and construction material requirements for expanding or rehabilitating airfields are usually similar to the requirements for new construction and ADR. Before using an existing facility for personnel, an inspection (ideally both an EBS and EHSA performed in conjunction with one another) should be done by environmental and preventive medicine personnel to ensure that Soldiers are not being exposed to existing environmental and occupational health hazards.

6-29. Upon completion of construction, the airfield manager or other individual authorized to monitor and control onsite aircraft operations can then certify the airfield and issue a notice to airmen (NOTAM) to change the airfield status.

AIRFIELD DAMAGE REPAIR

6-30. *ADR* **encompasses all actions required to repair airfield and landing zone operating surfaces and infrastructure or services to conduct operations at a base or location seized from the enemy or offered for use by a HN. It also includes repairs required to sustain operations or to reestablish operations after enemy attack at an airfield.** Airfields could be subjected to damage by an increasingly capable and complex array of destructive weapons, including cannon fire, rocket fire, small or large bombs, and bomblets. EH such as UXOs (to include scatterable mines and unexploded bomblets) and improvised explosive devices (IEDs), a variety of potential barriers, and other hindrances may challenge efforts to make airfields capable of supporting air traffic. Army engineers normally conduct minimal ADR as part of a forcible entry operation, focused on runway clearance and surface repair. ADR operations, however, include not only the actions required to rapidly repair airfield operating surfaces, but also the infrastructure to conduct operations at a base seized from the enemy or offered for use by a friendly HN. It also includes repairs required to sustain operations or to reestablish operations after enemy attack at either a FOB or main operating base (MOB).

6-31. Engineers must conduct a damage assessment, prepare for EH reconnaissance and removal, understand the repair quality criteria, and know the requirements for the minimum aircraft operating surface. Air Force technical experts and/or airborne RED HORSE elements may be included as a part of the Army combat engineer element participating in the forcible entry operation to approve the aircraft operating surface, control aircraft landing and departure, and serve as liaison to the airfield opening team. The airfield opening team will typically work with GE elements to take the airfield to a higher standard of repair after the lodgment area has been secured.

6-32. Pavement damage categories are shown in figure 6-1. Damage to the pavement includes both the apparent crater damage and the upheaval of pavement around the crater. The damage category for a given munition depends on the delivery method and extent of penetration and charge size. See UFC 3-270-07.

Damage Category	Probable Munitions	Probable Charge Size
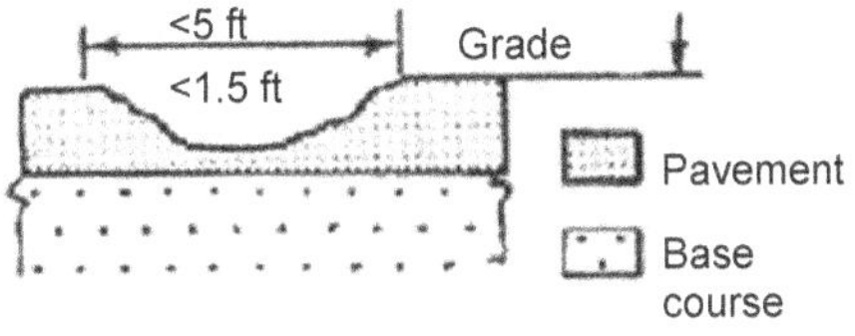 A. Spall/scab <5 ft, Grade, <1.5 ft, Pavement, Base course	Small rocket Cannon fire Contact-fused munitions	5-8 lb (2.3-3.6 kg)
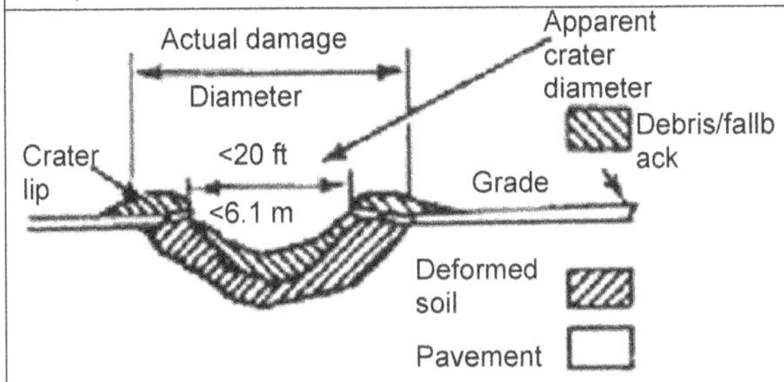 B. Small crater — Actual damage, Diameter, Apparent crater diameter, Crater lip, <20 ft, <6.1 m, Grade, Debris/fallback, Deformed soil, Pavement	Large rocket Clustered munitions Small concrete penetrators	5-35 lb (2.3-15.8 kg)
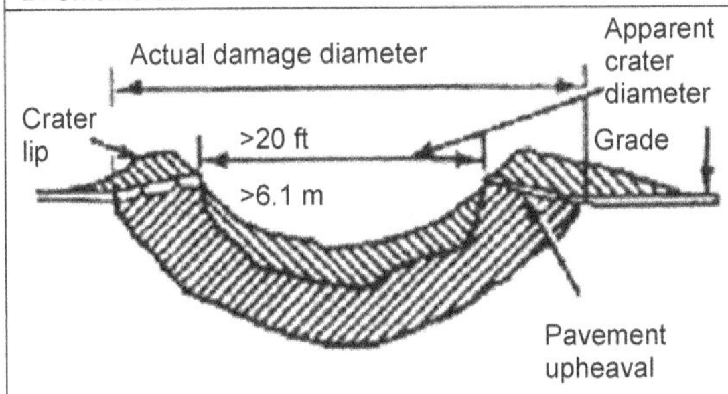 C. Large crater — Actual damage diameter, Apparent crater diameter, Crater lip, >20 ft, >6.1 m, Grade, Pavement upheaval	Bombs Delay-fused munitions Large concrete	<100 lb (45 kg)

Figure 6.1. Airfield damage categories

6-33. The airfield commander prioritizes essential ADR missions, usually in the following order:

- Reconnaissance or damage assessment.
- EOD.
- Minimum airfield operating surface (MAOS) repair.
- Repair to operational facilities, communication systems, ammunition storage facilities, essential maintenance facilities, fuel storage and distribution, utilities, and on-and-off base access routes as a result of indirect damage due to direct attack explosives that missed their primary targets. Environmental and occupational health hazards are included in these considerations.

6-34. Emergency repairs are conducted to provide a temporary fix. This allows the earliest possible resumption of air missions. The service that is responsible for the airfield determines the minimum operating strip and performs crater and surface repairs. The minimum operating strip is the minimum width and length required for an aircraft to land and take off. Normally, the largest area of the airfield with the least amount of damage is selected and identified as the minimum operating strip. All EH, including remotely delivered mines, must be cleared from the minimum operating strip before surface repair starts.

6-35. Army engineers organic to the BCTs will typically conduct the initial forcible-entry ADR. This is performed by airborne and air assault engineer elements centered on combat engineering skills and using specific force packages or modules to conduct emergency repairs using primarily a sand grid. At this level of repair this is a forward aviation combat engineering task enabling mobility operations. See FM 3-34.2 for a discussion of forward aviation combat engineering operations.

6-36. Air Force engineers have sole responsibility for conducting emergency repairs of established U.S. air bases. This is done by specific force packages or unit type codes (UTCs) formed from Air Force RED HORSE or Prime BEEF units. Currently, the Air Force uses the following three types of emergency repairs (depending on the nature of the damage):

- Crushed stone over debris.
- Choke-ballast repair.
- Choke ballast over debris.

6-37. Army engineers use the following two methods to make "beyond emergency" repairs to established U.S. air bases:

- Stone and grout repair.
- Concrete cap repair.

Air Force, Navy, and Marine engineers use similar techniques. The airfield commander directs the priority of pavement repair effort, allowing permanent repair to begin as soon as the tactical situation, available equipment, and labor permit. Pavements outside the minimum operating strip, including taxiways, usually have a lower repair priority. Deliberately marking and/or clearing EH (to include UXOs and IEDs) must be done before permanent repairs. Usually EOD personnel are available for these types of area clearance operations, but engineers may have to perform these tasks in their absence if time is critical and the risk is acceptable.

6-38. Army engineers are responsible for helping Air Force RED HORSE or Prime BEEF teams to repair critical air base support facilities, when such repairs exceed the Air Force's capability. Methods for repairing indirect damage are much the same as ordinary engineer construction techniques.

AIRFIELD MAINTENANCE

6-39. Airfield maintenance is the routine prevention and correction of damage and deterioration caused by normal use and exposure to the elements and aircraft traffic. Routine maintenance includes inspections; stockpiling materials for repair and maintenance work; maintenance and repair of pavement surfaces and drainage systems, dust and control; and snow, ice, and FOD removal. The procedures and considerations for airfield maintenance are similar to those for road maintenance and repair. The materials used for airfield maintenance are generally the same as those used for airfield construction and repair.

6-40. FOD removal is accomplished using motorized sweepers. The user of the airfield, in coordination with engineers responsible for airfield maintenance, should do routine FOD inspections.

6-41. Upon completion of an airfield repair or a maintenance mission, a repair evaluation must be conducted before resuming aircraft operations. When conducting the repair evaluation, the following should be considered:

- **Repair compaction.** The strength of the backfill, debris, or subgrade materials must be verified. Depending upon the repair method used, the thickness and strengths of all surface and/or base course materials must also be verified. The soil structure should be tested using a dynamic cone penetrometer (DCP) to determine the California bearing ratios (CBR) of each layer. These tests must be accomplished before placing the FOD covers, AM-2 matting, stone and grout, asphalt, concrete, or other surface materials that would prevent the use of the DCP.
- **Surface roughness.** The final grade of the repair must be checked using line-of-sight profile measurement stanchions, upheaval posts, or string lines to ensure that the repair meets the surface roughness criteria. In the case of a crushed stone repair without a FOD cover, the repair surface should be checked for loose aggregate or potential FOD.
- **Foreign object damage covers.** FOD covers should be no more than 5° off parallel with the runway centerline. Connection bolts are checked and all bolts are verified to be tight and secure

between panels. Anchor bolts are checked and all bolts are verified to be secure and the FOD cover is checked to ensure that it is held snugly against the pavement surface. In taxiway and apron applications, the leading and trailing edges of the FOD cover must be anchored. The side edges must also be anchored if the cover is located in an area where aircraft will be required to turn.

- **Setting and curing.** If the repairs are capped with concrete, stone and grout, or rapid-set materials, verify that the surface material has set and that adequate cure time is allowed before aircraft operations.

- **Clean up.** For all repair methods, verify that the areas are cleared of any excess repair materials.

- **Airfield certification.** The on-site engineer responsible for the repair will certify that the repair was accomplished according to the procedures in the appropriate UFC and other applicable publications. The repair procedures will be documented on an ADR log. This form will then be updated to reflect subsequent aircraft traffic and required maintenance throughout the history of the repair. If another team replaces the initial repair team, this form is given to the follow-on team. This information will be useful in planning or performing any further maintenance and/or upgrade of the repairs. Upon completion of repairs, the status of the airfield or repairs is provided to the airfield manager or other individual authorized to monitor and control on-site aircraft operations. This individual can then issue a NOTAM to change the airfield status.

This page intentionally left blank.

Chapter 7

Roads and Railroads

The line that connects an army with its base of supplies is the heel of Achilles–its most vital and vulnerable point.

John S. Mosby, Colonel, Confederate States of America (1887)

Maintaining forward-deployed forces during contingency operations requires an extensive logistics network. An adequate ground LOC network is a critical part of this network and one of the keys to sustainment operations in today's OE. Engineers are responsible for road and railroad construction, maintenance, and repair. Roads and railroads are built to meet mission and operational requirements. Depending on the METT-TC, they may form the primary LOCs during a contingency operation. From nation-building and humanitarian assistance, through combat operations, as unit AOs expand along both contiguous and noncontiguous lines due to technological advances and the natural evolution of modern warfare, roads and railroads will continue to be crucial to operational support throughout the AO.

ROAD CONSTRUCTION, MAINTENANCE, AND REPAIR RESPONSIBILITIES

7-1. The purpose of a military road is to allow traffic to get from one point to another quickly and efficiently, for as long as necessary, based on operational requirements. Roads are classified according to their degree of permanence and the characteristics of traffic they are designed to support. Sound engineering logic and the ever-changing, dynamic battlefield environment dictate that existing roads be used whenever possible. When suitable road networks do not exist or cannot be used, roads are upgraded or constructed to support operational requirements. Combat roads and trails are a combat engineering mission that is discussed in FM 3-34.2. Higher-level road work is a GE mission. The specific application of this level of road construction is discussed in FM 5-430-00-1.

7-2. Military roads are rarely constructed to meet the exacting standards of comparable civilian construction, to include environmental standards. Their degree of permanence varies depending on how long they are needed. Combat trails or earthen roads may be hastily cut pathways designed to initial standards to enhance mobility for only a short time (less than six months). More permanent road networks, such as MSRs and primary LOCs, are designed to temporary standards to sustain mobility for a longer period of time (up to two years). During contingency operations, nearly all roads are constructed to temporary standards. In some rare cases, semipermanent and permanent roads may be designed to provide long-term mobility (up to 20 years). Permanent roads are often done in conjunction with USACE or similar organizations and the employment of civilian contractors or the oversight of Service GE organizations.

7-3. Military roads are classified as either Type A, B, C, or D depending on the amount of traffic they are expected to sustain per day. Type A roads are designed for the highest capacity, while type D roads are designed for the lowest. Only road types B, C, and D apply to TO construction. See FM 5-430-00-1, chapter 9, table 9-1.

7-4. During contingency operations, most military roads that are constructed do not receive asphalt cement or portland cement concrete surface due to the time required and the added cost for this type of construction. These types of pavements may not be required to sustain heavy traffic over extended periods of time. Instead, most new military roads are surfaced with either sand, gravel, crushed rock, stabilized soil, or the best locally available materials. Selected high-use portions of the road may be surfaced to support this requirement. Gap-crossing sites will require special consideration. This allows for future upgrades and permits the maximum use of readily available materials to rapidly complete the road. Army engineer units have the following responsibilities as directed by the appropriate commander:

- Road reconnaissance.
- Maintenance, repair, and upgrade of existing roads.
- Construction of new roads.
- Recommendation of traffic pattern flow based on terrain and construction considerations.
- Topographic and other geospatial support.

ROAD CONSTRUCTION, MAINTENANCE, AND REPAIR PLANNING

7-5. The following factors must be considered when establishing any road network:

- **Mission.** Operational and mission requirements determine the minimum road classification and design requirements based on the expected period of usage and the anticipated traffic load.
- **Enemy.** Threat capabilities and anticipated types of action could impact the methods of construction and affect the road location and design. Creating choke points and other potential ambush points should be avoided when possible.
- **Terrain and weather.** The location of a road is dictated by operational and mission requirements. Using existing roads, when possible, is preferred to avoid unnecessary construction. Existing slopes, drainage, vegetation, soil properties, weather patterns, and other conditions may affect layout and construction.
- **Troops and support available.** Local materials, labor, and equipment are used whenever and wherever possible. Simple or preexisting designs, such as those in the TCMS that require a minimum of skilled labor and specialized equipment, are used whenever possible.
- **Time available.** Speed is critical to establishing a road network during a contingency operation because of the rapid and dynamic tempo of military operations. It is essential to save as much time as possible by efficiently using the minimum amount of resources. Effective project management techniques should be used. Good planning, careful estimating, sound scheduling, and thorough supervision speed job completion and save time, labor, equipment, and materials. Wherever possible, use staged construction to allow early use of roadways while further construction, maintenance, repairs, and upgrades continue.
- **Civil considerations.** Civilian property restrictions, existing structures, restricted areas, cultural beliefs, environmental considerations, and other factors may also affect road layout and construction.

7-6. One of the most critical parts in planning the establishment of a road network is site investigation. Site investigation includes reconnoitering the proposed routes and determining existing soil properties and terrain characteristics. This requires a thorough knowledge of soils engineering, hydrology, and engineer technical reconnaissance. A detailed site investigation will serve as the foundation behind the design of a new road and/or the upgrade, repair, and maintenance of an existing road.

7-7. Route reconnaissance (see FM 3-34.170) to evaluate the traffic-bearing capabilities and condition of existing roads supports route selection decisions and determines the improvements needed before a route can carry the proposed traffic. Route reconnaissance is classified as either hasty or deliberate. The way in which route reconnaissance is performed depends upon the amount of detail required, the time available, the terrain problems encountered, and the tactical situation. The composition of the reconnaissance element depends upon these protection considerations and a risk assessment by the commander directing the reconnaissance. Hasty route reconnaissance determines the immediate military trafficability of a specified route. It is limited to the critical terrain data necessary for route classification. The results are part of the mobility input to the common operational picture (COP). Information concerning the route is updated with additional reports as required by the situation and/or the commander's guidance. A deliberate route reconnaissance is conducted when enough time and qualified technical personnel are available. Deliberate route reconnaissance is usually conducted when operational requirements are anticipated to cause heavy, protracted use of the road and may be the first reconnaissance conducted or follow the conduct of a hasty route reconnaissance. When available, an automated route reconnaissance kit (ARRK) can provide engineer units with an automated reconnaissance package that allows the reconnaissance element to collect and process reconnaissance information (see appendix B). An overlay is made with attachments that describe all pertinent terrain features in detail. This overlay forms part of the mobility input to the COP and is maintained by the engineer unit tasked to perform the reconnaissance.

7-8. The engineer reconnaissance team is briefed as to the anticipated traffic (wheeled, tracked, or a combination) and the anticipated traffic flow. Single-flow traffic allows a column of vehicles to proceed while individual oncoming or overtaking vehicles pass at predetermined points. Double-flow traffic allows two columns of vehicles to proceed simultaneously in the same or in opposite directions. The reconnaissance team may also be asked to determine the grade and alignment, horizontal and vertical curve characteristics, and the nature and location of obstructions. Obstructions are defined as anything that reduces the road classification below what is required to support the proposed traffic efficiently. Obstructions include—

- Restricted lateral clearance, including traveled way width, such as bridges, built-up areas, rock falls or slide areas, tunnels, and wooded areas.
- Restricted overhead clearance, including overpasses, bridges, tunnels, wooded areas, and built-up areas.
- Sharp curves.
- Excessive gradients.
- Poor drainage.
- Snow blockage.
- Unstable foundations.
- Rough surface conditions.
- Other obstacles, including CBRN contamination, roadblocks, craters, EH (to include mines and other UXOs and IEDs), cultural sites, and environmental restrictions. Existing bridging may require special attention, as it is often a weak link. It may be necessary to conduct bridge reconnaissance and classification computations (see chapter 8).

ROAD CONSTRUCTION

7-9. Reconnaissance to support the construction of a new road may be classified as either area or route reconnaissance (see FM 3-90 and FM 3-34.170). Area reconnaissance is a search conducted over a wide area to find a general site suitable for construction. Route reconnaissance is an investigation of a particular site or an undeveloped, but potential, route. Before starting the actual reconnaissance, the engineer reconnaissance team should conduct a map reconnaissance of the site or area, to include studying aerial photographs; reviewing available geological and hydrological information and other geospatial information and considering any other available relevant information. The engineer reconnaissance team may request the following sources of information in planning reconnaissance missions and in making the preliminary study of a specific mission:

- Existing intelligence reports and threat analysis.
- Existing strategic and technical reports, studies, and summaries.
- Existing road, topographic, soil, vegetation, and geologic maps or any other geospatial information.
- Existing aerial reconnaissance reports.
- Existing road design information or maintenance plans.

7-10. An air and/or map reconnaissance includes a general study of the topography, drainage pattern, and vegetation. Construction problems, camouflage possibilities, and access routes should be identified. A route reconnaissance plan is developed by selecting the areas to investigate and the questions to be answered to support the reconnaissance.

Air and/or map reconnaissance can be used to eliminate unsuitable sites, but cannot be relied on for site selection. Digital imagery enhances the usefulness of this method of reconnaissance.

7-11. While air and map reconnaissance can effectively minimize needed ground reconnaissance, it cannot replace ground reconnaissance. It is on the ground that most questions must be answered, or that most observations tentatively made from available information are verified.

7-12. The engineer reconnaissance team may also determine soil properties on-site and at potential borrow pits and quarry sites along the proposed route. Soil properties, such as the liquid limit, plasticity index, CBR, and gradation are required to design a new road's pavement structure, or upgrade an existing road's pavement structure based on the anticipated traffic that the road will be required to support. These soil properties are also required to evaluate the suitability of aggregate taken from potential borrow pits and quarries for use in road construction, maintenance, and repair.

7-13. Site selection is a crucial step in new road construction. Future problems can be avoided by careful reconnaissance and wise consideration of future operational requirements. A project that is poorly laid out will not meet the requirements for construction ease and efficiency, maintainability, usability, capacity, and convenience.

7-14. Drainage patterns are also important in site selection. When the tactical situation permits, roads should be located on ridgelines. Thus, natural drainage features minimize the need for costly and time-consuming construction of drainage structures. Whenever possible, avoid subsurface water. If it is impossible to avoid road construction in saturated terrain, water tables must be lowered during construction. Steps must also be taken to minimize water's adverse effect on the strength of the supporting subgrade and base course.

7-15. Where possible, avoid obstacles, such as rivers, ravines, and canals to minimize the need for bridge construction or for other similar structures. Such construction is time-consuming and calls for materials that may be in short supply. Make maximum use of existing structures to decrease total work requirements. Do not bridge an obstacle more than once. Should gap crossing be necessary, ensure that the proper type of bridging or other method provides an adequate and sustainable solution.

7-16. These existing soil properties, grade and alignment, and horizontal and vertical curve characteristics are all considered when designing a road. To sustain traffic, roads must have a suitable pavement structure to support traffic loads and suitable geometric characteristics to allow traffic to rapidly and safely move along the route (see FM 5-430-00-1).

7-17. To sustain traffic, roads have a crowned driving surface and pavement structure, a shoulder area that slopes directly away from the driving surface to provide drainage off the driving surface, and side ditches for drainage away from the road itself (figure 7-1). The shoulder areas and side ditches along many roads may be minimal depending on their location and their road classification.

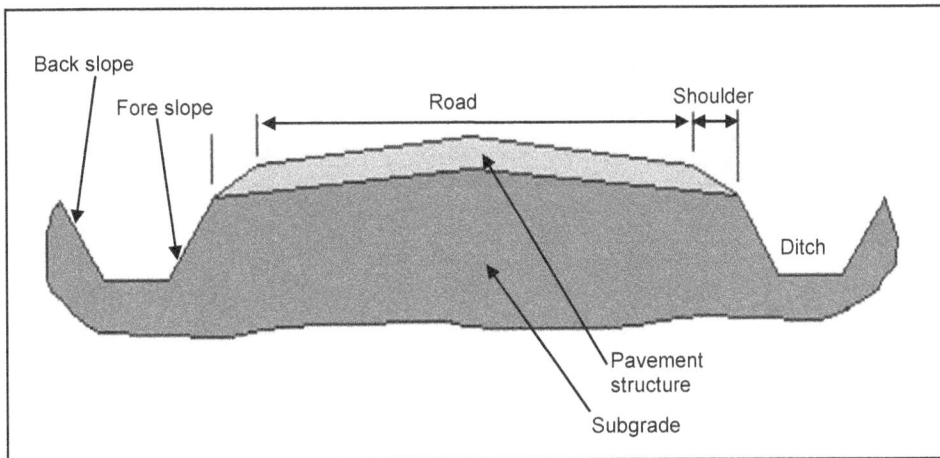

Figure 7-1. Typical road cross section

7-18. A road's pavement structure sits on top of the subgrade or the soil in place. A layer of compacted subgrade sits on top of the subgrade, and a layer of select material sits on top of the compacted subgrade. A layer of subbase material sits on top of the select material, and the base course sits on top of the subbase. A road may have a flexible pavement (asphalt) or a rigid pavement (concrete) surface on top of the base course (figure 7-2). The thickness of a layer depends on the strength of the layer below it. Depending on design requirements, some layers may not be required. These parts of the pavement structure distribute wheel loadings over a wide area within the pavement, thereby reducing pressures on the subgrade soils. During contingency operations, nearly all roads are constructed as aggregate-surfaced roads. These designs permit the maximum use of readily available materials and are easy to upgrade, permitting great flexibility to respond to changing operational and mission requirements. When possible, the roadbed should be aligned to take advantage of the most favorable surface and subsurface terrain. An alignment over soil with good properties meets the design standards for strength and stability and minimizes the need to remove undesirable materials.

Asphalt Pavement
Base
Subase
Select material
Compacted subgrade

Figure 7-2. Typical flexible pavement structure cross section

7-19. Traffic flow over roads is far more efficient if curves and grades are held to a minimum. Even gentle curves significantly decrease traffic capacity if there are too many on a route. Therefore, lay out all routes with a minimum of curves by making the tangent lines as long as possible. The availability of long tangents is influenced by terrain. It is also limited by other principles of efficient location, such as minimizing earthwork, avoiding excessive grades, and obtaining desirable soil characteristics.

7-20. Horizontal curves are circular curves. They connect tangent lines around obstacles, such as buildings, swamps, and lakes. The following four types of horizontal curves (figure 7-3, page 7-6) may be used in road construction:

- Simple.
- Compound.

- Reverse.
- Spiral.

Note. During contingency operations, only simple and compound curves are used because they are the simplest to design and construct.

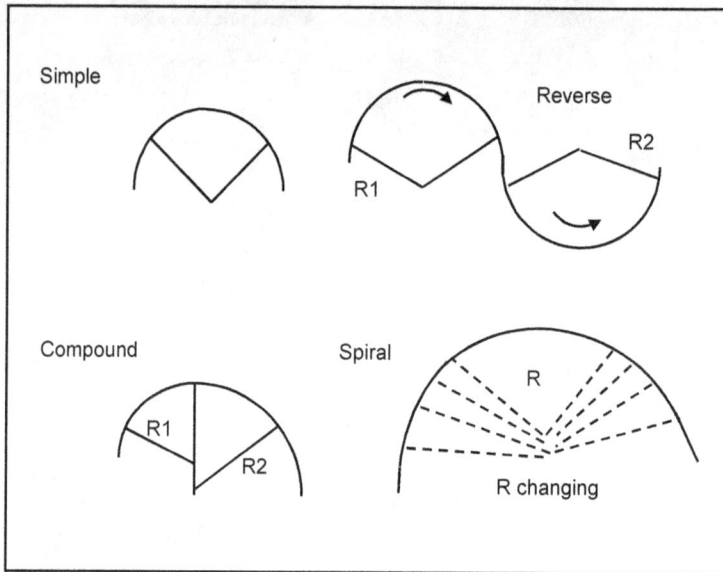

Figure 7-3. Horizontal curve types

7-21. Vertical curves are parabolic as opposed to circular horizontal curves. Vertical curves allow traffic to maintain reasonable speeds, provide adequate visibility for stopping and passing, and provide a smooth transition for increased comfort and safety. Overt and invert are the two types of vertical curves (figure 7-4). Vertical curve design determines the amount of cut and fill required to construct a road.

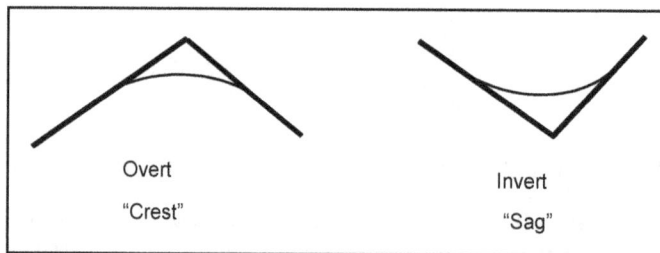

Figure 7-4. Vertical curve types

7-22. Earthmoving operations are usually the largest single work item on any project involving the construction of a road, unless the road will have significant gaps to cross. Any step that can be taken to avoid excessive earthwork will increase job efficiency. Since all roads are a series of grades that seldom appear in nature, it is inevitable that some earthwork must be done. However, the amount to be done should be minimized by properly locating the route. The engineer should take advantage of all prevailing grades that fall within the required specifications. Excessive grades should be avoided and steep hills should be bypassed whenever possible. If the route must negotiate excessively steep hills, it should run along the side of the hill. This may result in a longer route, but will prove to be more efficient in terms of earthwork and trafficability. Following contour lines on hillsides or ridgelines also avoids excessive grades and drainage construction. It is important to make a careful analysis of the geology and ground cover within the proposed area of construction. Avoid wooded areas, extremely rocky soils, undesirable humus, unnecessary clearing, and earthwork.

7-23. When there is need for both cutting and filling at various points along a project, use excavated material to construct embankments if possible. This reduces the need for earth handling. Plan balancing so that it fits the hauling capabilities of available equipment. Even though it is desirable to balance earthwork throughout a project, long hauling distances may make it more practical to open a nearby borrow pit to obtain fill material or to establish spoil areas to dispose of excess soil. Obviously, balancing cannot be done where excavated material cannot be used for embankment. If possible, roads should be located near construction materials. Readily available construction materials require less haul assets.

7-24. When a general route has been selected for new construction, a construction survey is initiated. In this survey, the team obtains data for all phases of construction activity. Surveys include reconnaissance, preliminary, final location, and construction layout.

7-25. A reconnaissance survey provides a basis for selecting feasible sites or routes and furnishes information for use in later surveys. The techniques discussed in the sections above on reconnaissance and site selection should be used. If a location cannot be selected on the basis of this survey, it will be chosen in the preliminary survey.

7-26. A preliminary survey is a detailed study of a location tentatively selected on the basis of reconnaissance, survey information, and recommendations. Surveyors run a traverse along a proposed route, record the topography, and plot results. Several surveys may be needed if reconnaissance shows that more than one route is feasible to meet the specified requirements for the road. If the best available route is not already chosen, it should be selected at this time.

7-27. A final location survey is conducted if time permits. Permanent benchmarks for vertical control and well-marked points for horizontal control are established. This enables construction elements to accurately locate and match specific design locations with those on-site.

7-28. A construction layout survey is the final operation before construction begins. In this instrument survey, alignment, grades, and locations are provided to guide construction operations. To enable construction to begin, exact placement of the centerline is made; curves are laid out; all remaining stakes are set, such as slope, grade, and shoulder; necessary structures are staked out; culvert sites are laid out; and other work is performed. This survey is used until construction is complete. The main purpose of construction surveys is to ease and control construction. The number of surveys conducted and the extent to which they are carried out are largely governed by available time, construction standard, and by personnel and material assets. Roads constructed to initial standards may be constructed with minimal preplanning and construction control. However, extensive surveys may be conducted for a road built to temporary standards. The quality and efficiency of construction is strongly related to the number and extent of surveys and other preplanning activities.

7-29. Adequate drainage is essential during construction of a military road or airfield. Immediately provide adequate drainage for the site to ensure that all water that might interfere with construction operations is removed. Eliminate construction delays and subgrade failures due to ponding of surface water by aggressive, timely development of a drainage system. Include temporary measures, such as pumping. During clearing and grubbing operations, keep existing or natural watercourses clear and fill and compact holes and depressions to grade. Rough crown and grade must be maintained to permit water from precipitation, sidehill seeps, and springs to move freely away from worksites by gravity flow. If water is permitted to pond, the subgrade becomes saturated and fails under load, earthmoving is impeded, and the need for equipment maintenance is increased.

7-30. In permanent peacetime construction, underground drains are often used because of efficient use of space, environmental considerations, and safety practices do not permit large open ditches, particularly for disposal of collected runoff. In contrast, designs for road drainage in contingency operations use surface ditching almost exclusively because of limited pipe supplies and the absence of storm sewer systems to collect runoff. The drainage system is designed to remove surface water effectively from operating areas, to intercept and dispose of runoff from adjoining areas, to intercept and remove detrimental conditions of the selected design storm, and to minimize the effects of exceptionally adverse weather conditions.

7-31. The proposed use of the road is considered. If it is to be used only for a short time, such as one or two weeks, a detailed drainage design is not justifiable (see FM 3-34.2). However, if improvement or expansion is anticipated, design drainage so that future construction does not overload ditches, culverts, and other drainage facilities. Drainage problems are greater when all-weather use occurs whether than when only intermittent use occurs. The availability of engineer resources should be considered. Heavy equipment (such as dozers, graders, scrapers, and excavators) is commonly used on drainage projects. But where unskilled labor and hand tools are readily available, much work can be done by hand. When the necessary reconnaissance and mission analysis are complete, the engineer prepares an estimate of the work and materials required and a plan for carrying out the construction. The engineer must schedule the priority and rate of construction and provide for the even flow of material to ensure orderly progress. Schedules must continually be updated to accommodate changed field conditions or other exigencies. In addition to their planning function, the schedules can also serve as progress charts.

7-32. When earthwork estimation, equipment scheduling, and necessary surveys are complete, the construction sequence can begin. The construction site is prepared by clearing, grubbing, and stripping. These operations are usually done with heavy engineer equipment. Hand or power felling equipment, explosives, and fire are used when applicable. The factors determining the methods to be used are the acreage to be cleared, the type and density of vegetation, the terrain's effect on equipment operation, the availability of equipment and personnel, and the time available for completion. For best results, a combination of methods should be used, choosing each method for the operation in which it is most effective.

7-33. Cut and fill operations are conducted when clearing, grubbing, and stripping are finished. Cut and fill operations are the biggest part of the earthwork in road construction. The goal of cut and fill work is to bring the route elevation to design specifications. Throughout the fill operation, the soil is compacted in layers (lifts). Compaction is achieved with self-propelled or towed rollers. The end product is a structure that minimizes settlement, increases shearing resistance, reduces seepage, and minimizes volume change. The advantages that accompany soil compaction make this process a standard procedure for constructing embankments, subgrades, and bases for road and airfield pavements. Cut and fill and compaction efforts are intended to achieve the final grade. This alignment takes into consideration super elevation along curves to ensure load stability, and falls within the grade specifications required for the military road. When final grade is achieved, ditching is cut to control drainage runoff and the road is crowned along its centerline. The road is now ready for surfacing.

7-34. During contingency operations, consider whether or not to pave a road by taking account of the urgency of its completion, the tactical situation, the expected traffic, the soil-bearing characteristics, the climate, the availability of materials and equipment, and the necessity of dust control. Pavements, including the surface and underlying courses, are divided into two broad types—rigid and flexible. The wearing surface of rigid pavement is made of portland cement concrete. Asphalt cement concrete pavements are classified as flexible pavements.

7-35. Flexible pavements are used almost exclusively in contingency operations. They are adaptable to almost any situation and fall within the construction capabilities of normal engineer troop units. Rigid pavements are not usually suited to construction requirements during contingency operations. Because flexible pavements reflect distortion and displacement from the subgrade upward to the surface course, their design must be based on complete and thorough investigations of subgrade conditions, borrow areas, and sources of select materials, subbase, and base materials.

7-36. Another option to improve the road's ability to support traffic is soil stabilization. The goals of soil stabilization are stabilization of expansive soils, soil waterproofing, and dust control. Strength improvement increases the load-carrying capability of the road. Dust control alleviates or eliminates dust generated by vehicle and aircraft operations. Soil waterproofing maintains the natural or constructed strength of a soil by preventing water from entering it. Stabilization is generally accomplished by either mechanical or chemical methods.

In mechanical stabilization, soils are blended and then compacted. In chemical stabilization, soil particles are bonded to form a more stable mass. Additives are used (such as lime, bitumen, or portland cement).

7-37. All unpaved roads will give off dust under traffic. Dust is usually an inherent problem. The amount of dust that an unpaved road produces varies greatly, depending on local climatic conditions and the quality and type of aggregate used to construct the road. Common dust control agents include chlorides, resins, natural clays, asphalts, and other commercial binders and membranes. Dust control and soil waterproofing can be carried out by applying these agents in a spray (soil penetrants); a mix (admix); or by laying aggregate, membrane, or mesh as a soil blanket. The agronomic method, using vegetation cover, is suited to stable situations, but is rarely useful during contingency operations. Ensure that the aggregate surface of the road has good gradation, there is a good crown on the driving surface and good drainage, equipment is calibrated accurately and working properly, and rehearse the application of the agent using a test strip for the effective application of any stabilizer or dust control agent.

7-38. Expedient surfaces may be used as a temporary means to quickly cross small areas with extremely poor soil conditions, such as swamps, quicksand, and wetlands when there is not enough time or resources for standard road construction. The following are the two types of expedient roads:

- **Hasty.** Hasty expedient roads are built quickly to last only a few days.
- **Heavy.** Heavy expedient roads are built to last until a durable standard road can be constructed.

Expedient surfacing methods include cross-country tracks, Army tracks, chespaling, corduroy, landing mat, chain-link wire mesh, plank roads, and snow or ice roads.

UPGRADING EXISTING ROADS

7-39. Wherever possible, existing facilities are used. In most areas, an extensive road network already exists. With expansion and rehabilitation of the roadway and preparation of adequate surfaces, this network can carry required traffic loads. Upgrading an existing road, combined with routine maintenance and repair, usually involves reducing or eliminating obstructions. It is the preferred method of improving the trafficability of a selected route. Techniques, equipment, and materials needed for upgrading are generally the same as those required for new road construction. A changing tactical situation and unpredictable military operations may also require that engineer troops modify and expand completed construction. The location of a road should allow for potential expansion. Expanding an existing route or facility conserves manpower and material and permits speedier completion of a usable roadway.

ROAD MAINTENANCE AND REPAIR

7-40. Road maintenance is the routine prevention and correction of damage and deterioration caused by normal use and exposure to the elements. Repair restores damage caused by abnormal use, accidents, hostile forces, and severe environmental actions. Rehabilitation restores roads that have not been in the hands of friendly forces and do not meet operational requirements.

7-41. Routine maintenance and repair operations include inspections, stockpiling materials for maintenance and repair work, maintenance and repair of road surfaces and drainage systems, dust and mud control, and snow and ice removal. The main purpose of maintenance and repair work is to keep road surfaces in usable and safe condition. It also increases route capacity and reduces vehicle maintenance requirements. Effective maintenance begins with a command-wide emphasis stressing good driving practices to reduce unnecessary damage. Once damage has occurred, prompt repair is vital. After deterioration or destruction of the road surface begins, rapid degeneration may follow. A minor maintenance job that has been postponed becomes a major repair effort involving reconstruction of the subgrade, base course, and roadway surface. The following principles should be observed in conducting sound road maintenance and repair:

- **Minimize interference with traffic.** To keep surfaces usable, maintenance and repair activities should interfere as little as possible with the normal flow of traffic. A temporary bypass may be required.

- **Correct the basic cause of surface failure.** Efforts spent to make surface repairs on a defective subgrade are wasted. Any maintenance or repair job should include an investigation to find the cause of the damage or deterioration. That cause must be remedied before the repair is made. To ignore the cause of the damage is to invite prompt reappearance of the damage.
- **Reconstruct the uniform surface.** Maintenance and repair of existing surfaces should conform as closely as possible to the original construction in strength and texture. Simplify maintenance by retaining uniformity. Spot-strengthening often creates differences in wear and traffic impacts, which are harmful to the adjoining surfaces.
- **Assign priorities.** Priority in making repairs depends on the operational requirements, commander's guidance, traffic volume, and hazards that would result from complete failure of the facility.

7-42. The purpose of maintenance inspections is to detect early evidence of defects before actual failure occurs. Frequent inspections and effective follow-up procedures prevent minor defects from becoming serious and causing major repair jobs. Surface and drainage systems and the road pavement should be inspected. Pavement surface defects can usually be attributed to excessive loads, inferior surfacing material, poor subgrade or base conditions, inadequate drainage, or a combination of these conditions. Ensure that all drainage channels and structures are unobstructed. Special vigilance must be exercised during rainy seasons, spring thaws, and after every heavy storm.

7-43. Generally, the materials required in road maintenance and repair are the same as those used in new construction. Maintenance activities may include opening pits and stockpiling sand and gravel, base materials, and premixed cold patching material. Materials should be placed in convenient locations and in sufficient quantities for emergency maintenance and repair. Stockpiles should be arranged for quick loading and transporting to key routes.

7-44. In some areas, extensive repairs are often needed to make roads usable. Advance engineer units usually do this work. Under the pressure of combat conditions, repairs are sometimes temporary and hurriedly made with the most readily available materials. Such expedient repairs are intended only to meet immediate minimum needs. As advance units move forward, other engineer units take over additional repair and maintenance. Early expedient repairs are supplemented or replaced by more permanent work. When surfaces are brought to a standard that will withstand the required use, only routine maintenance is performed.

7-45. Engineer units establish a patrol system to cover the road net for which the unit is responsible. Periodic patrols by other elements, such as the military police, who use the road net on a sustained and frequent basis, may also assist with this system. Engineer road maintenance and repair modules are organized with personnel, equipment, and supplies to accomplish road repair and maintenance in a specific area. As many modules as needed are organized to cover the total AOR. The traffic level and the limited durability of a road sometimes make it necessary to put the maintenance function on a 24-hour-a-day basis in heavy traffic areas. A squad-sized road maintenance and repair module equipped with a dump truck, grader, and hand tools can usually carry out all maintenance and minor repairs encountered on a 5- to 15-mile stretch of road. This module or force package can be increased or decreased, and more or fewer miles can be assigned to a module as the mission dictates. Security conditions may affect the size and composition of these elements and the method of employment.

7-46. During contingency operations, winter weather may present special maintenance problems. Regions of heavy snowfall require special equipment and material to keep pavements and other traffic areas open. Low temperatures cause icing on pavements and frost effects on subgrade structures. Alternate freezing and thawing may cause damage to surfaces and block drainage systems with ice. Spring thaws may cause both surface and subgrade failures and may damage bridging. Winter maintenance consists chiefly of removing snow and ice, sanding icy surfaces, erecting and maintaining snow fences, and keeping drainage systems free from obstruction. Each command should publish a comprehensive snow and ice control plan that clearly specifies the responsibilities of engineer and nonengineer units. Engineer and nonengineer patrols must be established to monitor snow and ice conditions within the AO. Available snow and ice control equipment and supplies must be allocated to support the plan.

RAILROAD RESPONSIBILITIES AND PLANNING

7-47. The ability to move troops and materiel may well decide the outcome of a conflict. Railroads provide one of the most effective and efficient forms of land transportation available to forces during contingency operations. They can move great tonnages of materiel and large numbers of personnel long distances. They move with considerable regularity and speed under practically all weather conditions. Railroads are flexible and versatile; rolling stock may be tailored for almost any use. Extensive railway systems exist in most regions of the world with an interoperability provided by standard equipment and common gauge. Due to these capabilities, railroads may be the preferred means of transportation during contingency operations. The degree to which a rail system may be exploited depends on its capacity (length and condition of existing track, condition of rolling stock, and other facilities) and its ability to operational support and mission requirements while still maintaining essential commercial traffic. FM 4-01.41 describes the organizations, processes, basic construction and maintenance standards, and systems involved in rail operations.

7-48. Engineers are responsible for new railroad construction. The transportation railway battalion is responsible for operating railroads and performing rehabilitation and routine maintenance. Each transportation railway battalion may be assigned from 90 to 150 miles of main line with terminal operating and maintenance facilities, signaling equipment, and interlocking facilities necessary for operation. Where HN agreements exist, day-to-day operations and maintenance may be largely conducted by the local work force.

7-49. Reconnaissance and selection of new routes is done by transportation units in coordination with engineers. Although transportation units have the responsibility for routine maintenance, engineers must be prepared to provide construction support in cases where additional maintenance beyond the organic capabilities of transportation units is required. All existing facilities must be used to the maximum extent possible to minimize construction time and effort. New railroad construction will normally consist of short spurs to connect existing networks with military terminals or to detour around severely damaged areas. The focus of engineer effort should be on modifying and repairing existing railroads to meet operational and mission requirements. Local labor and management are key to the rapid modification and continuing maintenance of existing facilities. Local personnel can often supply materials and skilled labor to speed the work and relieve military personnel for other projects. Local railway operating personnel are also a source of information on existing operations and supply facilities in a given area. Railroads constructed during contingency operations may have lower factors of safety, sharper curves, and steeper grades than recommended by the American Railway Engineering Association. Once the minimum standard for immediate service has been attained, phased improvements can be made, provided the importance of the line justifies the effort.

7-50. A simple steel stringer-type bridge supported on timber trestles or piles satisfies most railway bridging requirements. The construction process begins with a determination of design requirements. In most cases the design criteria should consider copper E80 loads (a train with a locomotive weight of 520 metric tons and axle loads equal to 37 metric tons). The engineer must establish immediate liaison with the Transportation Corps to develop the—

- Mission and required capacity of the proposed systems.
- Type and size of rolling stock to be operated.
- Track gauge.
- Initial, intermediate, and final terminal points along the route.
- Required servicing and maintenance of facilities.
- Connections with other rail systems.
- Required maximum gradient and degree of curvature.
- Scheduling or timetable for construction.
- Direction of future development and expansion.

7-51. Upon determining the design requirements, transportation unit representatives and engineers will conduct a reconnaissance to determine the siting of the rail system. The surveys, studies, and plans required for constructing a railroad are necessarily more elaborate than those for most road construction. Studies of the best available topographic maps, imagery, and other geospatial products narrow the choice of routes to be reconnoitered. Factors that affect the location of a route include logistics, length of line, curvature, gradients, and ease and speed of construction. Each of the factors of METT-TC may have an effect on the determination of the site of a railroad, just as they affect the location of a road.

7-52. Logistics are nearly always the primary consideration when selecting a rail route during contingency operations. Normally, a rail line will extend from a SPOD, APOD, beachhead, or other source of supplies in theater to the logistics support areas sustaining the forces present. Alternate routes are desirable for greater flexibility of movement and as insurance against cases of mainline obstruction as a result of threat actions, wrecks, washouts, floods, fires, or landslides.

7-53. The length of line (mileage from point of origin to terminus) is important only when it adds materially to the time of train movement. As much as a 30 percent increase in mileage is permissible when it proves advantageous to the other factors involved.

7-54. Curvatures should be minimized as much as possible and be consistent with the speed of construction. Curvature for a military railroad will depend largely on the maximum rigid wheelbase of cars and locomotives. Superelevation is used to counteract centrifugal force on curves by raising the outer rail higher than the inner.

7-55. The ruling grade of a route is the most demanding grade over which a maximum tonnage train can be handled by a single locomotive. Where diesel electric units are used, a single locomotive may consist of two or more units coupled to work as a single locomotive that is controlled from the cab of the leading unit. The ruling grade is not necessarily the maximum grade. Steeper grades can be negotiated with the use of an additional locomotive as a helper engine or, if the grade is very short, the train may be carried over the crest by momentum. Since military railroads operate at slow speeds, the ruling grade must be kept to a minimum. As always, the necessity for rapid construction must be a top priority.

7-56. The route should be chosen so that the railroad line can be quickly constructed using minimal resources. Transportation facilities must be available as soon as possible to support contingency operations. Many additional hours of earthwork and grading can be avoided by a careful route selection.

7-57. A complete ground reconnaissance of the possible routes is required. The reconnaissance team should note odometer and barometer observations of distances and elevations, general terrain characteristics, the controlling curvatures, soil and drainage conditions, bridge and tunnel sites, the size and type of bridges needed, intersections with railways or important roads, availability of ballast and other construction material, and points at which construction units would have access to the railway route. Factors to be taken into consideration include the roadbed, rock cuts, hillsides, drainage, security, water supply, passing track, and surveys.

7-58. The roadbed should be built on favorable soils. Clay beds, peat bogs, muck, and swampy areas are unstable foundations and provide unsuitable soils for building fills. Cuts through unfavorable soils will slough and slide. Seek minimum earthwork in locating the roadbed and track. Where rock cuts are proposed, bedding planes should dip away from the track to prevent rockslides. Locations at the foot of high bluffs subject the track to rock falls, slides, and washouts. Rock work is time-consuming and should be avoided whenever practicable. In the temperate zone, choose sites along the leeward side of hills. This prevents snowdrifts and resists the effect of winds.

7-59. The proposed site should facilitate drainage or prevent the need for it. Ridge routes are best for this purpose, but may be exposed to enemy fire or observation. Avoid locations that require heavy bridging. Note that diesel equipment cannot be operated over a track that is inundated above the top of the rail because water will damage traction motors. If a steam operation is planned, an adequate water supply must be available at 15- to 20-mile intervals along the route. Suitable sites for passing sidings must be planned. Passing track spacing depends on traffic density and expected peak conditions of traffic flow.

7-60. The preliminary survey includes cross sections along the feasible routes. Trail locations are plotted and adjusted to give the best balance of grades, compensated grades, cuts, and fills. This establishes or fixes the line of the railroad.

Field survey parties locate the precise line and stake it. This requires much more precision than the location survey of most new roads, since curves and super elevations must be accurately computed.

7-61. When the necessary reconnaissance and surveys are complete, the engineer prepares an estimate of the work and materials required and a plan for carrying out the construction. The engineer must schedule the priority and rate of construction and provide for the even flow of material to ensure orderly progress. Schedules must continually be updated to accommodate changed field conditions or other exigencies. In addition to their planning function, the schedules can also serve as progress charts.

RAILROAD CONSTRUCTION

7-62. As a first stage in organizing the work, the engineer divides the line into sections in which special features (such as bridges, stations, yards, and rock cuts) can be constructed while other work is in progress. Work can proceed concurrently at several locations. The standard construction sequence is as follows:

- Clear and grub.
- Prepare the subgrade by cutting or filling and compacting.
- Unload and distribute track materials.
- Align and space cross ties.
- Place line rails or ties.
- Place gauge rail on ties to ensure proper spacing.
- Line the track.
- Unload ballast.
- Raise and surface track.
- Make final alignment.

7-63. In addition to the actual rail line, certain facilities are necessary to rail operations or are required due to particular physical conditions. Sidings are auxiliary tracks next to the main line. They are used for meeting and passing trains, for separating and storing equipment that breaks down en route, and for storing rolling stock that cannot be moved to its destination. The transportation unit responsible for operating the railroad will determine siding locations in coordination with the engineers. Sidings are built parallel to the rail line. The siding should be 250 feet longer than the longest train that will use it. Generally, the siding has a turnout at either end.

7-64. Highway and road crossings at grade should be avoided wherever possible. When crossings must be installed, they should be constructed so that the axis of the road is approximately perpendicular to the centerline of the railroad. Rail crossings carry one track across another at grade and permit passing wheel flanges through opposing rails. The design of frogs to allow these crossings depends on the angle at which they cross. In military railroads, most frogs are made of precast, immobile rails that can be easily installed.

7-65. Wyes are used in place of turntables, which are normally impractical for use in contingency operations. Wyes may be installed at engine terminals, summits, junctions, and railheads as time permits. In some cases, the wye's stem may be long enough to permit turnaround of the entire train.

7-66. Service facilities should be laid out so that servicing operations can be performed in the proper sequence as the locomotive moves through the terminal. The usual relation of operations and facilities from terminal entrance to terminal exit is—

- Inspection (inspection pits or platforms).
- Lubrication (during inspection) (oil and grease service areas).
- Cleaning fires and ash pits (for steam locomotives).
- Coal, sand, diesel oil, and water (appropriate facilities).
- Running repairs (engine house).
- Outbound movement (the ready track and wye).

7-67. Structures are needed for crew headquarters, maintenance personnel, tools, material storage, and block stations. Block stations are facilities that house the switching and signaling equipment that controls train movements.

7-68. A railhead is at the end of a railroad line. Yards are a system of tracks that serve the following three basic functions:

- One or more tracks long enough to receive an entire train.
- A system of shorter tracks for the storage or classification of freight.
- Departure tracks on which rolling stock from the classification yard may be assembled for dispatch.

7-69. In addition to the auxiliary facilities described above, other specific construction requirements may be dictated by the terrain or operational requirements. Special equipment, materials, and expertise may be required to construct a railroad and its accompanying facilities to quickly and efficiently support units.

RAILROAD MAINTENANCE AND REPAIR

7-70. Rail lines and supporting facilities must be inspected regularly to ensure adequate maintenance and proper operation. Necessary action must be undertaken as quickly as possible to minimize future repair requirements. Preventive maintenance, including the proper cleaning and lubrication of equipment and machinery, will minimize the need for unnecessary maintenance and repairs. The upkeep of railroads is essential to the smooth flow of troops and supplies to the needed areas. Railroads are susceptible to maintenance problems, and vulnerable to enemy attack, guerrilla operations, and sabotage. Railroads used by the transportation railway service are normally already located and constructed. The rail transportation officer's task is to make the most efficient use of existing facilities by maximizing maintenance efforts.

Chapter 8

Bridging

History shows that army campaigns in undeveloped countries have often involved waging war against natural obstacles, rather than against a foe.

Air Marshall E. J. Kingston-McCloughry

Military traffic engaged in rapid decisive maneuvers must be able to cross wet or dry gaps in existing road networks or natural high-speed avenues of approach. Very few LOCs will exist without some form of bridge, bypass, or detour. Maneuver forces and logistics support depend on four types of bridging—tactical, support, LOC, and existing or permanent bridges. Tactical operations of combined arms forces within the BCT are primarily focused on the first two of these forms of bridging or the seizure of existing or permanent bridges. Tactical bridging is typically linked to combat engineers and immediate support of combined arms ground maneuver. See FM 3-90.12 for a more in-depth discussion of bridging as a component of combined arms gap-crossing operations. See FM 3-34.170 for the specifics of reconnaissance associated with bridging. Engineers enable mobility through construction, repair, and reinforcement of bridges; by providing bridge reconnaissance and classification; and in the construction of bypasses and detours. The specific mission undertaken is planned in a manner that maintains the momentum of the force. Bypasses and fording sites can be used to overcome obstacles when it is more feasible or when no bridges are available. Existing bridges may need to be repaired or reinforced to keep MSRs and LOCs open. As the tactical situation changes, MSRs are moved or adjusted to support the force. Forward elements may demand that expedient, standard, and nonstandard structures be emplaced to replace tactical bridging and support bridging and those assets returned for their use by the combat maneuver elements of the force. These types of bridges are not designed for the multiple passes that are typical for MSRs and will need to ultimately be replaced by other bridging. Requirements for engineer units to employ tactical, support, and LOC bridging will continue throughout the fight.

BRIDGE TYPES AND CATEGORIES

8-1. The two basic bridging types are standard and nonstandard (figure 8-1, page 8-2). While the two types could be combined as a hybrid of some nature, the bridge will normally be identified by the predominant components of the bridge. Standard bridging includes any bridging derived from manufactured bridge systems and components that are designed to be transportable, easily constructed, and reused. Examples of standard bridging include the Wolverine, dry support bridge (DSB), and Bailey bridges. Nonstandard bridging is purposely designed for a particular gap and typically built using COTS or locally available materials. They are used when time permits, and materials and construction resources are readily available; when standard bridging is inadequate, unavailable, or being reserved for other crossings; and the situation allows for unique construction. These bridges are left on-site, even when they are no longer necessary to support military movement. Nonstandard bridging is typically constructed by construction engineers or contractors using construction materials, such as steel, concrete, and/or timber.

8-2. There are three bridging categories (figure 8-1) and they are broadly defined by their intended purposes. These categories include tactical bridging, support bridging, and LOC bridging. The bridging category is typically dictated by the OE, gap characteristics, and equipment available. They are subordinate to the bridging types and can be standard or nonstandard. As the situation changes, crossing sites may be abandoned, improved, or replaced with the appropriate alternatives that provide appropriate solutions for each site.

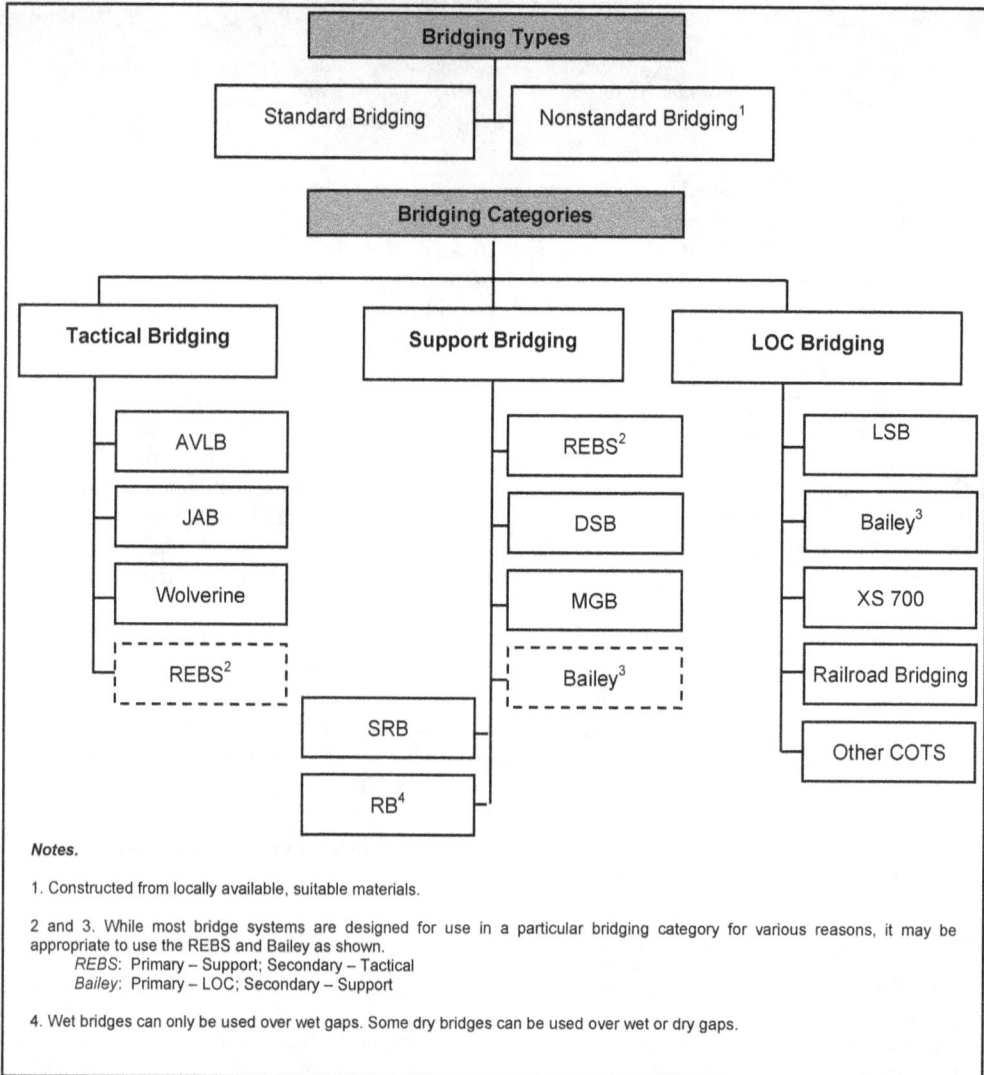

Figure 8-1. Types and categories of bridging

TACTICAL BRIDGING

8-3. Tactical bridging is rapidly deployable with the mobility to maintain the pace of operations with the maneuver force it supports. Tactical bridging is typically linked to combat engineers and immediate support of combined arms ground maneuver. The actual bridge can be deployed and recovered without exposing the crew to direct or indirect fire. There are little to no requirements for bank preparation when using tactical bridging assets. It takes minimal time to deploy and recover for temporary crossings. Engineers use primarily four bridge systems to conduct tactical bridging operations. They are the—

- Armored vehicle-launched bridge (AVLB).
- Joint assault bridge (JAB).
- Wolverine.
- Rapidly emplaced bridge system (REBS).

8-4. Unfortunately, the AVLB is an old system that is prone to maintenance problems, and lacks the speed to maintain momentum with heavy brigade combat team (HBCT) and Stryker brigade combat team (SBCT)-based maneuver forces. The REBS meets all the requirements for tactical bridging except that the crew is exposed to enemy fires during emplacement and can only support a military load classification (MLC) of 30. At this time, the Wolverine is the only bridge that meets all the desired tactical bridging criteria. Although tactical bridging can be used on LOCs, planners will often find that this limited resource should only be used for a limited time until some other bridging method can be employed. Tactical bridging is not designed to accept the repeated number of crossings associated with LOC sustained use. Additionally, it is often important to release these bridges so that adequate tactical bridging is available to support combat maneuver within the force and sustain the desired or required tempo of operations.

SUPPORT BRIDGING

8-5. Support bridging is used to establish semipermanent or permanent support to planned movements and road networks. It is used to replace tactical bridging as it provides greater gap-crossing capability to the force than tactical bridging. Units typically deploy and recover these systems when and where little or no direct fire threat exists. Bank preparation and improvement are important planning factors for support bridging. The support bridging category contains all float bridges in the Army inventory: the standard ribbon bridge (SRB) and the improved ribbon bridge (IRB). Other support bridges include the medium girder bridge (MGB), DSB, Bailey panel bridge, and REBS. Although a REBS is often considered a tactical bridge, it is more accurately described as a support bridge because it lacks crew survivability.

FLOAT BRIDGING

8-6. Float bridges are designed to cross maneuver forces over wet gaps by either building raft configurations to transport forces across the wet gap, or by emplacing bays to span the entire width of the wet gap. There is generally no design limit to the length of this bridge. The normal limiting factor is the quantity of bays and boats; however, velocity or current of the water, tidal variation, water depth, underwater obstructions and floating debris, and entry and exit bank slopes can limit float bridge operations. Descriptions and construction techniques for the SRB are found in FM 5-34 and for the IRB the techniques are explained in TM 5-5420-278-10. Float bridging may be used when there is a lack of existing fixed facilities or no suitable construction materials to fabricate, reinforce, or repair existing bridges. When the situation calls for prolonged use or heavy traffic, an existing bridge should be upgraded or new construction initiated.

8-7. Criteria for establishing a float bridge site may be the same as those for general bridge site selection criteria. The following are specific float bridge site considerations:

- Banks should be low, firm, moderately sloping, and free from obstructions. Existing or easily prepared assembly sites are desirable.
- Stable banks should have a slope of 8° or less and a water depth of at least 48 inches on the nearshore.
- Water velocity near the shore should be less than 5 feet per second, if the current is faster (up to 10 feet per second), then additional boats and time will be required to emplace the bridge.
- Natural holdfasts for anchorages are desirable. Float bridging must be installed far enough downstream from a demolished or under-capacity bridge to avoid interference with reconstruction or reinforcement operations. Unstable portions of a demolished bridge and other debris that may damage the float bridge should be removed before the emplacement of the float bridge.

LINE OF COMMUNICATIONS BRIDGING

8-8. LOC bridging is generally conducted in areas free from the direct influence of enemy action. Typically, its primary purpose is to facilitate sustainment of the force. It can be used as a semipermanent or permanent structure. LOC bridges are built with the assumption that once emplaced, they will not be removed until a permanent structure is constructed to replace them. LOC bridges may be tactical fixed bridges if the intention is to leave the system in place for an extended period of time and they are not required for the support of combat maneuver. Planning factors should then account for the extended use of the bridge and any wear that will occur as a result of the extended use. A common consideration for LOC bridges is planning for the possibility that an existing permanent bridge has been damaged or is not strong enough for mission requirements. Engineers will repair and reinforce a bridge, if necessary, by either using standard or nonstandard materials to meet mission requirements. New construction of LOC bridges is possible; however, improving existing structures is the primary focus because of the intense resource requirements associated with new construction.

8-9. Railroad brides are LOC bridges and are classified as nonstandard bridges. The U.S. Army does not currently have design criteria for nonstandard railroad bridges nor does it maintain railroad float bridge equipment. Many varieties of standard railroad bridges are available through AFCS. Construction details and BOMs are given in TM 5-302. Standard railroad bridging is available for the Bailey bridge and certain contracted panel bridges. Repair and reinforcement of existing railroad bridges is a much more viable option than new construction in most cases. Nonstandard railroad bridging can be repaired or improved using any available and suitable materials. Railroad bridges will require specialized construction equipment and large quantities of labor. This generally precludes the construction of railroad bridges at locations away from existing rail lines. When a site must be selected, use the basic criteria for general bridge sites.

8-10. The urgency of the situation or lack of additional bridging assets may require that a railroad bridge be converted into a highway bridge by constructing a smooth roadway surface. The use of the bridge by both rail, wheeled, and tracked vehicles can be achieved by constructing planking along the ties between and outside the rails up to the level of the top of the rail. The roadway surface is made flush with the top of the rail with adequate distance from the rails to allow use by train wheels. The additional dead load of roadway decking must be factored into the bridge classification to determine safe traffic loads. Since railroad loadings are usually heavier than highway loadings, it is seldom practical to convert a highway bridge to railway use.

BRIDGE SITE SELECTION

GEOSPATIAL CONSIDERATIONS

8-11. Engineers have the ability to use the engineer function of geospatial engineering to greatly improve situational understanding (to include terrain) and select optimal bridging sites. High-resolution satellite imagery or UAS video are precise pictures of terrain. The requirement for the engineer is to have the appropriate software. Engineer terrain teams should assist in determining conditions in areas at or around potential gap-crossing sites. Terrain teams have software that can assist in mission planning by determining soil conditions, hydrology, vegetation types, general weather patterns, and other useful aspects of the terrain.

RECONNAISSANCE

8-12. Engineer reconnaissance teams should be used to collect data to determine acceptable terrain and conditions for new construction. Using the results of reconnaissance (see FM 3-34.170), planners can determine which type of bridge or bridge combinations are right for the mission based on available resources. The location ultimately chosen for the bridge is determined by numerous factors which are reflected in its structural design. Primary screening considerations include—

- **Access and approach roads.** Determine if the preexisting roads are adequate. Remember that the time to construct approaches can be a controlling factor in determining if a crossing site is feasible. Ensure that approaches are straight, with two lanes, and less than a 6 percent slope.
- **Width.** Determine the width of the gap to be spanned at both normal and flood stage for wet gaps.
- **Banks.** Estimate the character and shape of the banks accurately enough to establish abutment positions. Ensure that the banks are firm and level to limit the need for extensive grading. Select straight reaches to avoid scour.
- **Flow characteristics.** Determine the stream velocity and erosion data, taking into consideration the rise and fall of the water. Remember that a good site has steady current that runs parallel to the bank at less than 3 feet per second.
- **Stream bottom characteristics.** Record the characteristics of the bottom to help in determining the type of supports and footings required. Remember that an actual soil sample is useful in the planning process, particularly in wide gaps that may require an intermediate pier.
- **Elevation.** Determine and record accurate cross-sectional dimensions of the site for determining the bridge's height. Ensure that planners know of any existing structures that the bridge must cross over.
- **Materials.** Determine the accessibility of material for improving bank conditions, such as rock.

8-13. If these primary considerations appear favorable, planners may apply the following evaluation criteria:

- Proper concealment for personnel and equipment on both sides of the gap.
- The location of bivouac and preconstruction storage sites.
- Firm banks with less than a 5 percent grade to reduce preparation work. Less than 1 percent grade will also require site preparation.
- Terrain that permits rapid construction of short approach roads to existing road networks on both sides of the gap.
- Turnarounds for construction equipment.
- Large trees or other holdfasts near the banks for fastening anchor cables and guylines.
- A steady, moderate current that is parallel to the bank.
- A bottom that is free of snags, rocks, and shoals and is firm enough to permit some type of spread footing.
- Determination of the number of assembly sites for floating portions of the bridge, either upstream or downstream. If the current is strong, locate all assembly sites upstream from the bridge site.
- Proper siting of logistics sustainment operations to mitigate the possible effects of flooding.

EXISTING BRIDGES

8-14. Part of site selection is reconnaissance of existing structures to evaluate the physical details of existing bridges. Engineer reconnaissance teams inspect the bridge to determine its load-carrying capacity (classification) and its structural integrity. Engineer reconnaissance teams should determine whether the situation warrants emplacing a tactical, support, or LOC bridge. When a damaged bridge is being considered for repair or replacement, reconnaissance information should include a report on the serviceability of the structural members in-place, local materials that might be reused in other construction, and the potential for overbridging (see FM 3-90.12 and FM 3-34.343). Maximum use should be made of existing bridge sites to take advantage of the existing roads, abutments, piers, and spans that are serviceable.

8-15. Bridge reconnaissance is classified as either hasty or deliberate, depending on the amount of detail required, time available, and security in the AO. Both types of reconnaissance are fully discussed in FM 3-34.170. A deliberate reconnaissance is usually conducted in support of the MSR and LOC bridging operations since greater traffic requirements dictate that time and qualified personnel be made available to support the task. Use of the ARRK will assist the engineer reconnaissance team by tracking the location, speed, curve, and slope of roads and obstacles encountered along the route (see appendix B). An engineer light dive team can assist with the deliberate reconnaissance by providing nearshore and farshore crossing site data. Additionally, they can mark and prepare landing sites, riverbanks, and exit routes for the crossing force. A deliberate reconnaissance includes a thorough structural analysis; report on approaches to the bridge site; report on the nature of the crossing site, abutments, intermediate supports, and bridge structure; repair and demolition information; and the possibility of alternate crossing sites.

8-16. After proper reconnaissance, a bridge study is completed. This is the detailed analysis of the selected site. To complete a study, the engineer should—

- Request a topographic map to a scale of about 1:25,000. Use this map to plot the location and obtain distances and elevations for design purposes.
- Determine whether physical characteristics at the site limit normal construction methods or interfere with construction plant installation.
- Make a detailed survey to furnish accurate information from which the bridge layout can be developed, materials requisitioned, and the construction procedure outlined. Submit the survey as plan and profile site drawings.
- Conduct a foundation investigation. Develop a soil profile along the proposed bridge centerline and at pier and abutment locations (see FM 5-410).

BRIDGE CLASSIFICATION

8-17. An efficient MSR network must be capable of carrying all expected traffic loads. Often, bridging is the weak link in the load-carrying capacity of a route. Military standard bridging is assembled in modules that result in a bridge of known capacities. Support bridging is designed to pass an uninterrupted flow of combat and tactical vehicles that generally fall within a MLC below 60. However, some combinations of vehicles may exceed a given bridge design capacity. Where heavier loads are anticipated, it is best to designate MSRs along routes that already possess bridges with appropriate classification ratings, or to design and emplace bridges that can carry these loads. Selective use of fords in conjunction with MSR bridge sites may also provide a solution in selected cases.

8-18. Situations arise when it will be impossible to safely accommodate all traffic designated to cross MSR bridges. Guidelines are set for special crossings (caution and risk) for oversized or overweight loads on military standard fixed and float bridging. Specific guidance for determining special crossing is contained in FM 5-34, FM 5-277, TM 5-5420-212-10-1, TM 5-5420-278-10, and TM 5-5420-279-10. JTF engineer planners must recommend appropriate circumstances for risk or caution crossings to the commander and receive the delegation of authority for approval of such crossings if necessary. An engineer officer must periodically inspect the bridge for signs of failure when routine caution crossings are made and after each risk crossing. Structurally damaged parts must be replaced, repaired, or reinforced before traffic can resume. If necessary, an engineer light dive team can assist in determining the extent of any subsurface damage, and completing repairs

8-19. Not all civilian bridges are designed to support military MSR traffic and all required load classifications may not be known when forces initially enter the AO. There are an infinite number of types of bridges that forces may encounter in a given AO, and there is no single, easy approach to classifying all of them. Figure 8-2 depicts many of the various types of bridges that units may encounter in an AO. Table 8-1, page 8-8, gives the span construction types when recording them on the bridge reconnaissance report.

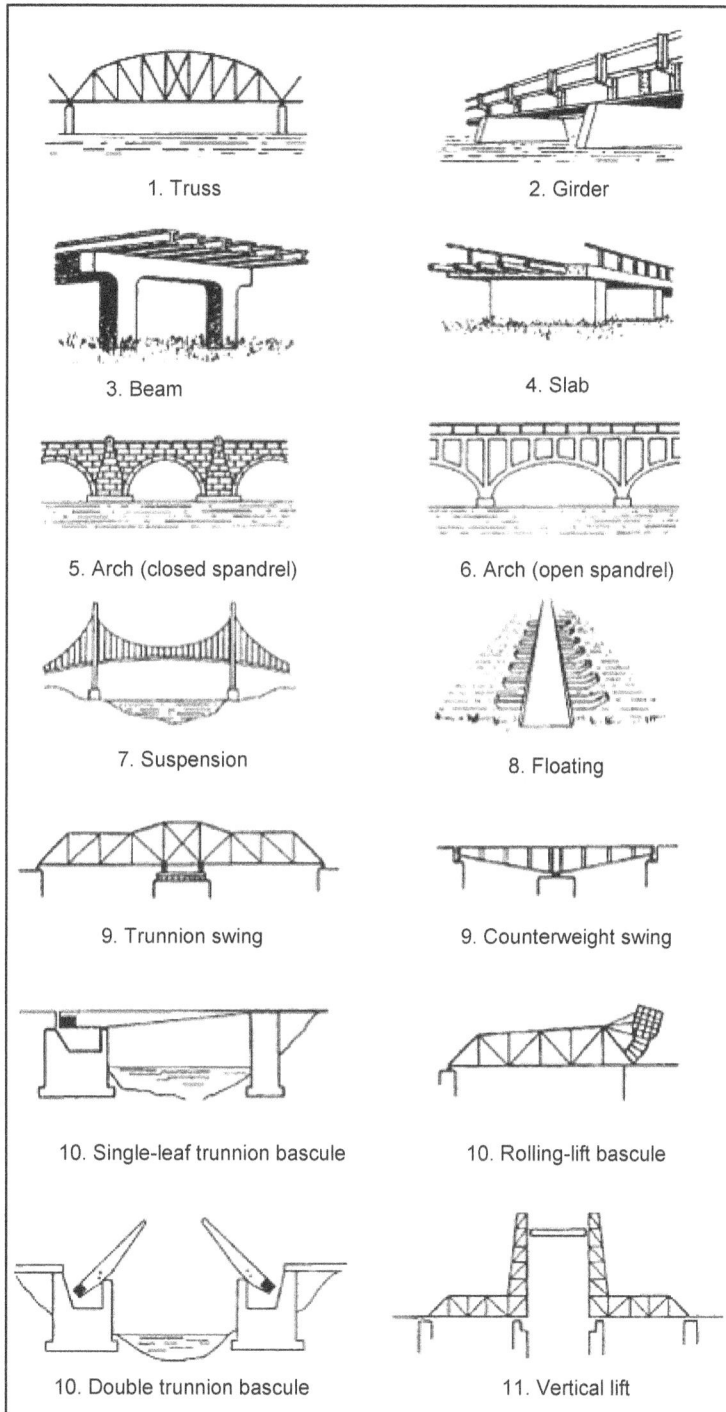

1. Truss

2. Girder

3. Beam

4. Slab

5. Arch (closed spandrel)

6. Arch (open spandrel)

7. Suspension

8. Floating

9. Trunnion swing

9. Counterweight swing

10. Single-leaf trunnion bascule

10. Rolling-lift bascule

10. Double trunnion bascule

11. Vertical lift

Figure 8-2. Selected bridge types

Table 8-1. Span Construction Types

Span Type	Number
Truss	1
Girder (including steel multigirder and two girder spans)	2
Beam (including reinforced or prestressed concrete and steel box beam spans)	3
Slab	4
Arch (closed spandrel)	5
Arch (open spandrel)	6
Suspension	7
Floating	8
Swing	9 (specify type in additional information)
Bascule	10 (specify type in additional information)
Vertical lift	11
Other	12 (specify type in additional information)

METHODS

8-20. Bridges are classified by either analytical or expedient methods. Careful analysis must often follow expedient classification. The situation and available time and information determine the method chosen. An analytical classification may be required if the bridge is of great importance. An engineer's estimate may suffice if similar bridges in the area have a known classification.

8-21. Bridge classification data can usually be found with the engineer unit in the AO containing a particular bridge. This unit is responsible for the area where the bridge is located along with the supporting geospatial products and imagery. If military engineers constructed the bridge, the design class or as-built plans should be on file. Satellite imagery and engineer intelligence studies often provide some level of bridge classification information for most of the potential AOs in foreign countries. If this is not the case for a specific bridge, engineer reconnaissance data will be used to classify it. The most reliable source of bridge classification information for civilian-constructed bridges is often the local civilian authorities. In most cases, complete design specifications, as-built plans, and the types and strengths of materials used in civilian bridges are available. Local, state, and county officials in the United States and in friendly foreign countries often impose maximum load limits or maximum permissible stresses on their bridges. It is important that these officials be consulted to determine maximum MLC that can be applied to the bridge in peacetime or for maneuver purposes. Corrosion and normal wear and tear tend to diminish a bridge's load-carrying capacity over time. The most recent evaluation of the bridge is desirable and currency of the evaluation is important to note. Based upon the engineer's evaluation of civilian reports, additional appraisal of a bridge's classification may be required. Correlation curves have been developed for some standard United States and foreign civilian-made bridges that relate known civilian bridge design loads to MLCs. These curves, discussed in FM 3-34.343, are generally useful for establishing a temporary bridge classification. The analytical method of classification is always preferred when time and information are available.

BRIDGE CLASSIFICATION AND MARKING

8-22. Bridge classification and marking is an engineer responsibility. It must follow classification systems established by STANAG 2010, STANAG 2021, and QSTAG 180, which permit the use of bridges at their maximum safe military capacities. The classification systems relate bridge capacity to the overall loading effect a vehicle might impose on a bridge. The responsible engineer organization in the area will classify bridges of military significance by the analytical method if possible. If the responsible engineer judges a posted temporary class accurate, the classification can be posted as permanent. Engineer units should keep records on each significant bridge within their assigned area.

EXISTING BRIDGE REINFORCEMENT AND REPAIR

8-23. Civilian bridges in the AO often suffer damage or may be below the load-carrying capacity required for use on an MSR or LOC. These bridges can be reinforced or repaired by engineers. Bridge reinforcement is the process of increasing the structure's load-carrying capacities by adding materials to strengthen the component parts, or by reducing span length. Bridge repair, on the other hand, means restoring a damaged bridge to its original load-carrying capacity or higher. Reinforcement or repair of existing bridges or sites has many advantages, but primarily those of economy of time and material. Since existing bridges are usually located on established routes, they will require less work on approaches and speed the flow of traffic. The availability of preexisting serviceable bridge components, particularly abutments and piers, conserves both time and materials.

BRIDGE REINFORCEMENT

8-24. Once the decision has been made to reinforce a bridge, several construction factors must be taken into consideration before detailed planning and execution are undertaken. Among these factors are details of the site, available materials, and possible construction methods. Pertinent questions concerning the site include—

- What parts of the original structure are still usable?
- What is the type of bridge and what are the span lengths?
- What are the characteristics of the waterway, particularly as to the use of additional bents or pile piers?
- Will the present approaches be satisfactory for a reinforced bridge?
- Will the intermediate supports and abutments also need to be reinforced? Are alternate sites available?

8-25. Materials that may be used include standard steel (preferred because of quality and speed of construction), stock timbers, other military items of issue, and local materials of adequate quality. Possible construction methods depend upon items of equipment available, working locations, and the nature of the repairs. A detailed discussion of bridge reinforcement can be found in chapter 4 of FM 3-34.343.

8-26. Another option for bridge reinforcement is the use of overbridging. Overbridging is a method used to reinforce, provide emergency repair, or augment existing bridges or bridge spans using standard bridging. It can be used in a variety of gap-crossing situations, but is typically used when time is critical and/or construction assets and resources are not readily available to make the existing bridge reliable. The inherent characteristics of each of these tactical bridges, including the fact that they do not require a gap for emplacement (zero gap bridge), make them a viable option for placement in support of close combat and other operations. Risk should be evaluated, however, when using these bridges to repair or replace damaged spans if the bridge will not be supported by a pier or abutment. In other situations when time is not as critical, enemy contact is less likely, support or LOC bridging is readily available, and/or the gap is beyond the span length of tactical bridging, the MGB, logistics support bridge (LSB), Bailey, Acrow®, or similar systems may provide an appropriate alternative. When considering this option, the repair or reinforcement of piers or abutments may be necessary; however, reduce the overall work effort by replacing a span or spans with support or LOC bridging.

BRIDGE REPAIR

8-27. Emergency repairs are usually governed by the requirement that a crossing site be available as soon as possible. Immediate need dictates the desired capacity and permanence of the structure. Where possible, standard bridging should be used to expedite repairs, and tactical bridging is designed for this purpose. In the absence of tactical or standard bridging, expedient methods may satisfy the requirements. Most emergency structures will later be reinforced, replaced, or rehabilitated. Bridge structures and surroundings, the nature of bridge damage, and the methods of repair are all so varied that no single preferred method will always be the preferred method. Experience with several methods will usually suggest a practical method of repair. Unless there has been an opportunity for advanced planning, the selection of repair methods should be left to the engineer commander who is responsible for the repairs. The factors upon which the engineer will base choices include the—

- Type of bridge.
- Nature of damage.
- Tactical situation and bridge requirements.
- Nature of the surroundings and immediately usable bypasses or detours.
- Troops and equipment available.
- Standard stock bridging materials and accessories available and the time involved in procurement.
- Local materials available.
- Time estimated for bridge repair versus time estimated for a detour or preparation of a bypass.

DETOURS AND BYPASSES

8-28. Detours and bypasses are important in ensuring continued movement of forces. The problem of crossing wet gaps is reduced considerably when an alternate route is available to a serviceable bridge, or to a bridge that can be repaired under favorable circumstances. Detours are usually of the following types:

- Alternate routing over other existing bridges which have not been damaged.
- Alternate routing over bridges with lesser damage, or routing to other locations.
- Alternate routing of highways over railroad bridges.

8-29. Bypasses are usually of the following types:

- Bypasses with a grade crossing around the bridge.
- Fords.
- Local ferries, rafts, or barges.
- Ice bridges in extremely cold climates.

8-30. When determining construction methods, consider the condition of existing roads and the approaches that connect with the detours and bypasses. The work necessary to make roads usable may outweigh the advantages of using these alternatives. Traffic-supporting properties, grade and alignment, built-up areas, and sharp curves or corners involving clearances are also important factors to ensure that vehicle requirements are met.

PART THREE

Other Sustainment Operations

Part three of this manual discusses how the GE function supports sustainment operations other than LOC support. Sustainment operations at any echelon are those that enable decisive and shaping operations to occur without pause by providing sustainment functions, security, movement control, terrain management, and infrastructure development. Although joint and Army engineers may not have the lead in conducting sustainment operations, they clearly have an important role in ensuring that they are successful. Part three discusses several very important engineer missions that facilitate these sustainment operations other than LOC support. Chapter 9 discusses protection construction support, chapter 10 discusses procurement of construction materials, chapters 11 and 12 discuss facilities, chapter 13 discusses real estate, chapter 14 discusses power generation, chapter 15 discusses pipelines, and chapter 16 discusses wells and water distribution. Each chapter provides the overall concept for executing a specific type of GE mission that supports sustainment operations.

Chapter 9

General Engineering Support to Protection

The art of war teaches us to rely not on the likelihood of the enemy not coming, but on our readiness to receive him; not on the chance of his not attacking, but rather on the fact that we have made our position unassailable.

Sun Tzu

In full spectrum operations, commanders must constantly and actively protect the force. The protection warfighting function is the related tasks and systems that preserve the force so the commander can apply maximum combat power. Preserving the force includes protecting personnel (combatant and noncombatant), physical assets, and information of the United States and multinational partners. The protection warfighting function has nine capability tasks that define it: safety; fratricide avoidance; survivability; air missile defense; AT; chemical, biological, radiological, nuclear, and high-yield explosives (CBRNE) defense; information protection; FHP; and operational area security. When synchronized with the other warfighting functions, protection will ensure that the force maintains the ability to fight and win (see FM 3-0). An integral part of protection success is GE support. In terms of protection, GE is focused on survivability and related AT tasks. Planners should use this chapter in conjunction with FM 5-103 and the TM 5-301 series which contain detailed information applicable to survivability and the broader protection warfighting function. Both documents are focused on survivability and AT and the subordinate specific focus area of hardening as performed by engineers or units supported by engineer technical expertise and equipment. Hardening is the act of using natural or man-made materials to protect personnel, equipment, or facilities.

THREAT

9-1. Because protection encompasses the entire operational spectrum (to include CONUS-based operations), the threat to U.S. forces varies extremely in nature and includes opponents who possess a wide variety of capabilities. Deployed Army forces may encounter various threats or be specifically targeted by terrorists, paramilitary elements, or even conventional military forces. All of these groups may seek to manipulate events and realize their desired goals by striking at U.S. forces. Commanders must commit significant resources to lower risk to acceptable levels. Adversaries who seek to destabilize an area will go to great lengths to expel U.S. forces and advance their agendas.

CONVENTIONAL THREAT

9-2.   The conventional threat that the United States became accustomed to and focused on during the Cold War still has aspects that remain, and the skill sets required to meet this threat still remain. However, during current and future operations, U.S. forces are more likely to face an unconventional enemy and encounter or work with nations of widely diverse political systems, economic capabilities, cultures, and militaries. Future challenges will involve the full spectrum of warfare against an evolving and asymmetric enemy. These challenges may range from conventional war to AT operations and insurgencies. U.S. forces may still face Soviet style weaponry and tactics, or they may be limited to low-intensity operations against nontraditional enemy forces. When facing a more conventional enemy force refer to FM 7-100 as a baseline document.

UNCONVENTIONAL/ASYMMETRIC THREAT

9-3.   The concept of asymmetric warfare is critical to understanding the OE. Asymmetry is a condition of ideological, cultural, technological, or military imbalance that exists when there is a disparity in comparative strengths and weaknesses. In the context of the OE, asymmetry means an adaptive approach to avoid or counter U.S. strengths without attempting to oppose them directly, while seeking to exploit U.S. weaknesses. The asymmetric approach is not a new phenomenon, but given the position and capabilities of the United States as opposed to its potential enemies, it is more likely to be used against the United States and its allies by other nations and nonstate adversaries. Potential opponents will seek to avoid United States strengths while exploiting perceived United States weaknesses.

9-4.   Various countries and nonstate entities have studied how the U.S. fights and have begun to devise ways to fight a technologically superior force and win. Future wars will not be fought within the confines of the conflicts of the past. Since it is difficult to predict who the next enemy will be, the United States does not always have the luxury of having studied these nations or nonstate adversaries. Therefore, the United States must be prepared to refocus quickly, learn fast, and rapidly apply lessons learned in training. Flexibility and initiative are key to being able to adapt doctrine, organization, training, materiel, leadership and education, personnel, and facilities (DOTMLPF) domain solutions to meet the challenges of a given adversary and the asymmetric approaches that may be applied against friendly forces.

PROTECTION CONSIDERATIONS

9-5.   The overarching implementing concept for the protection warfighting function is contained in the Army keystone protection manual. As the proponent manual, it describes how the Army applies resources to support this warfighting function and establishes the foundation of protection that will assist commanders in preserving combat power. In addition, FM 5-103 and the TM 5-301 series of manuals provide illustrations and design criteria for many of the examples provided in this chapter. Most of these illustrations are focused on survivability and AT and the hardening measures provided by engineering support.

9-6. Commanders are responsible for their unit's protection plan. Engineer involvement is critical in protection planning from two perspectives. They prepare their unit's protection plans and they provide input (and capability) to the units they support. As with other missions, engineer protection planning must be well thought out, logical, and integrated with other staff planning. Protection plans or policies must be developed in line with the command estimate process. IPB is an integral part of protection planning and serves as the basis for developing a protection strategy. IPB includes both threat analysis and the commander's assessment of unit vulnerabilities. Engineers must be involved in the IPB process to ensure that engineer intelligence needs are integrated into the reconnaissance and surveillance plan.

9-7. In developing a protection policy, commanders and their staff analyze the situation using the following process:

- They determine the composition of assets (personnel, equipment, and facilities).
- They define the threat and attack probability.
- They determine levels of protection for each asset.
- They identify constraints.
- They design protective systems to counter threats.

9-8. The engineer must ensure that the maneuver staff and commander develop an AT and protection policy based on the threat. The plan must balance the attack probability, the consequences of inadequate protection, and the cost of adequate protection (risk level). Planners use FM 5-480 to assist in the process of balancing the consequences and costs of protection and recommending appropriate levels of risk to the commander. The commander must set the priority of protection for his forces and equipment, local assets, infrastructure, and the local populace.

PROTECTIVE MEASURES AND TECHNIQUES

9-9. As experienced in Iraq and other locations, engineers are tasked to plan and execute infrastructure repair and FOB physical security, such as bunkers, guard towers, overhead covers, entry control points (ECPs) and facilities, protective berms, and vehicle checkpoints. Because of organic equipment and expertise, GE support is critical in supporting AT and protection.

DEFENSIVE MEASURES

9-10. Engineers play a significant role in protecting deployed U.S. forces. They have the capability, when given time, priority, and a thorough IPB, to effectively establish defensive measures to protect forces, facilities, and equipment from potential aggressors. The list of measures below will enhance a force's survivability and AT posture. The specific options that the engineer planner selects will be based on the—

- Specific threat in the AO.
- Degree of protection required.
- Time available.
- Materials available.

9-11. Basic considerations include—

- Eliminating potential hiding places near facilities.
- Providing an unobstructed view around all facilities.
- Siting facilities within view of other occupied facilities.
- Locating assets stored on-site but outside facilities within view of occupied rooms of the facilities.
- Minimizing the need for signs or other indications of asset locations.
- Minimizing exterior signs that may indicate the location of assets.
- Providing a 170-foot minimum facility separation from installation boundaries.
- Eliminating lines of approach perpendicular to buildings.
- Minimizing vehicle and personnel ECPs.

- Eliminating parking beneath facilities.
- Locating parking as far from facilities as practical.
- Illuminating building exteriors or exterior sites where assets are located.
- Securing access to power and/or heat plants, gas mains, water supplies, and electrical service.
- Locating public parking areas within view of occupied rooms or facilities.
- Considering minimum recommended separation of facilities or developing mitigating procedures if minimums cannot be met.
- Locating construction staging areas away from asset locations.
- Locating the facilities away from natural or man-made vantage points.
- Locating the facilities' critical assets within areas that do not have exterior walls when possible.
- Minimizing window areas.
- Covering windows next to doors, so that aggressors cannot unlock the doors through them.
- Securing exposed exterior ladders and fire escapes.
- Designing building layout so that there are no areas hidden from view from control points or occupied spaces.
- Arranging building interiors to eliminate hiding places.
- Locating assets in spaces occupied 24 hours a day, when possible.
- Locating activities with large visitor populations away from protected assets when possible.
- Locating protected assets in controlled areas where they are visible to more than one person. Placing mail rooms on the perimeter of facilities.
- Providing emergency back up power generation for critical activities and facilities.

9-12. Listed below are some of the potential missions where engineers may be tasked to construct protective devices or facilities to support protection. Depending on the time, availability, resources, and extent of the required protection, horizontal and vertical companies and/or other more specialized teams or sections may be assigned the mission. Planners should review FM 5-103 and consult TCMS for detailed protection criteria.

- C2 sites.
- Support facilities.
- Logistics sites.
- Troop concentration areas.
- ECPs.
- Vehicle checkpoints.

9-13. Another area where engineers contribute significantly to protection is with the engineer mine dog detachment. This detachment consists of dog handler teams equipped to conduct mine detection operations in support of the movement of troops within a TO. They support the force by detecting casualty-producing devices during route and area clearance operations, route reconnaissance, and other missions. See FM 3-34.2 and ST 20-23-8 for additional information.

AREA DAMAGE CONTROL AND INCIDENT MANAGEMENT

9-14. *ADC* measures are taken before, during, or after hostile action or natural or man-made disasters to reduce the probability of damage and to minimize its effects. ADC measures are typically applicable to offensive, defensive, and stability operations. ADC functions and missions are primarily referred to as GE missions that support many aspects of protection. While there are certain ADC measures that can be taken before and during a hostile action or disaster, the focus of the GE effort is in repair and reestablishment of operations. During offensive, defensive, and stability operations, ADC efforts normally occur in rear (or relatively secure) areas and are planned as part of rear area or base defense.

9-15. ADC operations can be extensive and involve various other Army branches, other Services, governmental and nongovernmental agencies, and HN assets. During a stability operation, the land component commander (LCC), ASCC, or USAID will coordinate the effort, depending on the mission and its associated C2 structure. If ADC operations occur during combat operations (offense and defense) it is an operational matter. If possible, engineers should consider the use of HN assistance, as it can be a vital resource, and also serves the purpose of putting those in the area of the disaster back to work in a meaningful fashion. This allows those personnel to become a part of the solution to the destruction rather than remaining part of the problems caused by the destruction. Early HNS identification and coordination are essential if they are to supplement the ADC effort. Responsibilities and support from the HN are normally negotiated at the theater level and as a part of the SOFAs and treaties.

9-16. Incident management is a national comprehensive approach to preventing, preparing for, responding to, and recovering from terrorist attacks, major disasters, and other emergencies. Incident management includes measures and activities performed at the local, state, and national levels and includes both crisis and CM (see JP 1-02). Regardless of whether DOD is conducting incident management as a part of homeland defense or civil support operations, military forces always remain under the control of the established 10 USC, 32 USC, or state active duty military chain of command. DOD is the lead, supported by other agencies, in defending against traditional external threats/aggression (such as air and missile attack). However, against internal asymmetric, nontraditional threats (such as terrorism), DOD may be in support of the DHS. When ordered to conduct homeland defense operations within U.S. territory, DOD will coordinate closely with other federal agencies or departments (see JP 3-26). Except for homeland defense missions, DOD serves in a supporting role for domestic incident management.

9-17. As a subset of incident management, CM are those actions taken to maintain or restore essential services and manage and mitigate problems resulting from disasters and catastrophes, including natural, manmade, or terrorist incidents. Responses requiring CM occur under the primary jurisdiction of the affected state and local government, and the Federal government provides assistance when required. When situations are beyond the capability of the state, the governor requests federal assistance through the President. The President may also direct the Federal government to provide supplemental assistance to state and local governments to alleviate the suffering and damage resulting from disasters or emergencies. The DHS or FEMA has the primary responsibility for coordination of federal CM assistance to state and local governments.

This page intentionally left blank.

Chapter 10

Procurement and Production of Construction Materials

Class IV stocks should be robust and ready for crisis projects. If engineers don't stock Class IV, no one else will.

S-4, 130th Engineer Brigade, Operation Iraqi Freedom After-Action Review

GE missions can be and usually are resource intensive. One of the key and often limiting resources is construction materials. Determining the method of construction and obtaining materials on time and in the quantity and quality needed must be synchronized to support the assembly of other resources (time, personnel, and equipment) to complete the project. Construction of any kind will fail if the required materials (or suitable substitutes) are not available. Efforts to obtain the proper material at the time, in the quantity, and of the quality needed must begin early during the planning phase (receipt of the mission or construction directive) and do not really end until the project completion and turnover. For procurement, engineers have the options of obtaining materials from CONUS through the service supply system, from countries as adjacent to the AOR as possible, and locally. Each method has inherent costs and benefits. Engineer units may be used, or a contractor hired, to produce the necessary materials. Whatever the method, obtaining resources must be an integral part of all planning and execution tasks to properly accomplish the mission.

METHODS OF CONSTRUCTION

10-1. In almost every contingency operation conducted, base camps and FOBs have been constructed throughout the AOs to support all aspects of the U.S. military mission. Living quarters, dining and recreation facilities, post exchanges (PXs), and a multitude of other support facilities are an important component of base camps and occupy significant space within a camp. There are multiple options for the construction of the supporting infrastructure for these facilities, ranging from using preexisting structures; erecting tentage; assembling pre-engineered metal or fabric buildings; bringing in modular buildings, trailer units, assembled prefabricated buildings, or manufactured buildings; or by constructing wood, steel, or concrete masonry unit (CMU) framed and supported buildings. Each of these methods has advantages and disadvantages in TO construction.

PREEXISTING STRUCTURES

10-2. Using preexisting structures (figure 10-1, page 10-2) can be the least time-consuming for providing needed facilities for mission accomplishment. The original design use, intended use, and current structural integrity are key factors in determining the feasibility of a preexisting structure that meets mission requirements.

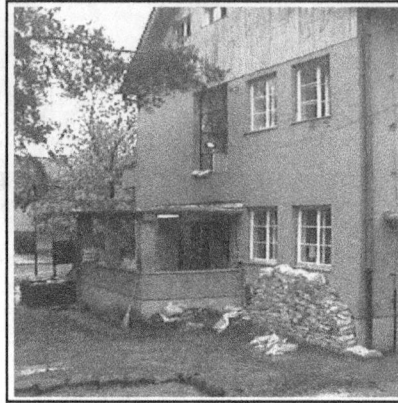

Figure 10-1. Preexisting structure

TENTAGE

10-3. The use of organic unit tentage (figure 10-2 and figure 10-3) or assembled packaged kits like "force provider" is another option. This option is quick in establishing basic life support areas and minor mission support areas (such as company, battalion, brigade, and division tactical operation centers/headquarters) if the tentage is already on hand and available; however, the impacts of long-term use of tents and the impact on the quality of life, and the level of personnel protection must be weighed in light of the mission requirements. The planned length of use must also be weighed against the ability to reuse the tentage later. The longer tentage is used and exposed to environmental conditions, the less likely it is to be easily repacked and stored for reuse. Also, as tentage is converted from tier 1 to tier 2, or tier 3, its reuse is also impacted. Consider the following:

- Is this only the initial (or immediate) standard?
- What is the plan for transition to a more enduring temporary or semipermanent facility?
- Is sufficient billeting capacity designed to allow for transitions between units?

Figure 10-2. General purpose (GP) medium tentage with wood floor

Figure 10-3. Tentage protected with HESCO Baston® revetments

PRE-ENGINEERED METAL OR FABRIC BUILDINGS

10-4. Pre-engineered metal or fabric buildings are structures that are completely assembled on-site out of standard components and materials brought to the site. They range from custom designs, to fit specific sites and usage requirements, to prepackaged and assembled kits ready for construction. Some of the advantages of pre-engineered metal or fabric buildings are—

- Rapidly constructed.
- Mobility/transportability of the materials.
- Flexibility of designs.
- Durability and low maintenance requirements.
- Minimal or no foundation preparation or requirements.

10-5. A disadvantage of some manufacturing techniques is that some of the major structural components are quite large and bulky in nature, making mobility/transportability a significant consideration. Other manufacturers have worked through that issue by fabricating panels or the major structural members on-site.

Pre-Engineered Metal Buildings

10-6. An example of a pre-engineered metal building system (assembled on-site) (figure 10-4, page 10-4) is the next generation of the K-Span® structure from M.I.C. Industries, Incorporated, known as the Ultimate Building Machine® (UBM®). The UBM, a self-contained manufacturing factory on wheels, is capable of fabricating and assembling an entire building at the construction site, producing a durable steel building in days. It is capable of being transported directly to the construction site via truck or airplane to anywhere in the world. A small crew of 10 to 12 workers can manufacture and assemble a 10,000 square feet structure in as little as a single day. The structures produced from various designs are unique and site specific with ground-to-ground, self-supporting panels that require no beams, trusses, columns, nuts, bolts, fasteners, screws, or sealants and they are virtually maintenance free. With the UBM system, units could build a *variety of facilities using these small crews and small quantities of building materials. A comparison of traditional TO construction to steel building construction using this technology, conducted by the U.S. Army Corps of Engineers, found these steel buildings require only half as much labor, 40 percent as much material, half as much construction time, and less than one-fourth the cargo space for materials transport. Although not as important as the above considerations, the steel buildings can be as much as 60 percent cheaper than wood TO structures.

Figure 10-4. Metal buildings constructed with the UBM in a contingency environment

PRE-ENGINEERED FABRIC BUILDINGS

10-7. The pre-engineered fabric building system is commonly referred to as giant tents or tent buildings. They are actually engineered fabric structures developed according to the prevailing construction/engineering standards that are prefabricated, customizable, and modular in design that can meet a wide range of military and civilian applications. Many manufacturers use in-house steel and fabric production so they have complete control over the manufacturing process, enabling stringent quality monitoring and lower cost operations. These systems typically require simple foundation preparation and allow for quick erections. Most are relocatable structures that are deployable wherever needed, with the ability to be recovered and then stored after use for the next requirement. With many of these systems, a small structure can be erected in hours by a small crew of 3 to 8 personnel aided by a lift truck with access platforms. The same structure, over time, could see use by multiple owners, in different locations, and be used for very different applications than the structure was originally supplied for. (An example would be a clamshell originally purchased for use as a dining facility in one contingency operation and finding use as a maintenance bay or as surge housing capacity in its next use.) These types of structures are available in multiple spans (ranging typically from about 20 to 80 meters wide by just about any length). These structures are adaptable to a multitude of purposes ranging from housing, logistics support warehousing requirements, to quality-of-life uses. Typically the fabric membrane will have about a 20-year service life expectancy depending on the extremes of the environments they are used in.

PRE-ENGINEERED TENSION FABRIC BUILDING SYSTEM EXAMPLES

10-8. Examples of pre-engineered tension fabric building systems are the medium shelter system (MSS) by Universal Fabric Structures (UFS)®, commonly known as a clamshell shelter and structures produced by a multitude of other venders like Rubb Building Systems™. Some examples of these tension fabric buildings are shown below. Figure 10-5 shows UFS MSSs. Figure 10-6 shows a genuine fabric structure. Figure 10-7 shows some of the products by Rubb Building Systems.

Figure 10-5. Clamshell structure

Figure 10-6. Rubb fabric structure

Figure 10-7. Tension fabric structures located at Balad Air Base, Iraq

MODULAR BUILDINGS OR TRAILER UNITS

10-9. Modular buildings or trailer units are types of facilities that are fabricated/assembled off-site, transported to the site, and placed in position. Modular buildings are manufactured in a controlled factory environment. The building or trailer is then delivered to a prepared building site for installation. A modular building comes complete with all the necessary components, including walls, floor trusses, windows, heating and cooling, plumbing, electrical wiring, and interior finishes. These structures vary in size and cost and can be very versatile. Modular buildings may be used for many purposes; they can be free standing or built inside an existing structure. Many modular buildings are constructed from steel. Modular buildings provide flexibility and other advantages over site-construction, such as—

- Cost savings.
- Speed of occupancy.

- Factory-controlled quality.
- Ease of expansion.
- Ease of relocation.

10-10.   The most significant advantage of modular buildings is the significant on-site time that is saved as opposed to on-site construction. Access and availability of an appropriate foundation are possible concerns/disadvantages when placing these structures on site. Figure 10-8 shows container buildings.

Figure 10-8. Containers used as life support areas at Camp Demi, Bosnia

PREFABRICATED OR MANUFACTURED BUILDINGS

10-11.   Prefabricated or manufactured buildings are types that consist of several factory built components that are assembled on-site to complete the unit. Prefabrication is used for many types of constructions because it saves time on the construction site. This can be vital to the success of projects where on-site construction time is limited or where weather (or the tactical) conditions may only allow for brief periods of construction effort on-site. Prefabricated building components are manufactured in a controlled factory environment. The components are then delivered to prepared building sites for assembly and installation. A prefabricated building comes complete with all the necessary components, including walls, floor trusses, windows, heating and cooling, plumbing, electrical wiring, and interior finishes. Prefabricated buildings can be used for many purposes; they may be freestanding or built inside an existing structure. Like modular buildings, many prefabricated building components are also constructed from steel.

10-12.   There are multiple advantages to prefabricated buildings; they are self-supporting and ready-made components are used, so the need for shuttering and scaffolding is greatly reduced. Overall construction time is reduced and buildings are completed sooner, allowing an earlier occupancy. On-site construction is minimized. Quality can be controlled while the components are in production. Typically less waste is generated during the production process and on-site. Molds for the different components can be used several times. There are also disadvantages; careful handling of prefabricated components, such as concrete panels, is required. Leaks can form at joints in prefabricated components. Transportation costs can be high, depending on the distance between the factory, the construction site, and the method of construction used. Figure 10-9 shows a manufactured building that was constructed by the U.S. Army for use by the New York Police Department after September 11 (or 9-11).

Figure 10-9. Manufactured building

ON-SITE CONSTRUCTION USING WOOD, STEEL, OR CONCRETE MASONRY UNITS FRAMING (TRADITIONAL CONSTRUCTION)

10-13. On-site construction using wood, steel, or CMU framing (traditional construction) standard basic construction techniques are used. One advantage is the flexibility of the designs. They can be modified to fit the existing site conditions perfectly and have personnel protective enhancements built directly into the plan. A significant disadvantage is the amount of time and availability of personnel to design and construct individual facilities in this manner on such a large scale. Figure 10-10 shows SEAhut cluster buildings and figure 10-11, page 10-8, shows a CMU structure.

Figure 10-10. SEAhut cluster

Figure 10-11. CMU constructed fire station

PROCUREMENT OF CONSTRUCTION MATERIALS

10-14.   Units may obtain GE construction materials using standard supply procedures that unify the way in which they are requested, managed, and distributed. Most construction materials are Class IV and their distribution in the AO occurs as depicted in figure 10-12. Note that this figure depicts a contiguous AO and may be much different than that encountered in a particular contingency. Many Class IV materials are also used for field fortifications, fighting positions, and other sorts of protection work, making it likely that they are in high demand and necessitating engineer involvement in distribution decisions. Class IV supplies are not maintained in significant quantities and are bulky. This makes handling and transportation over strategic distances difficult. Because of this, obtaining GE materials through normal supply channels is considered the least efficient and desirable method for GE missions. Engineers should only use this method after determining the materials are unavailable locally, that the proper quantity and quality cannot be met locally, or the cost to obtain them in this fashion is prohibitively high. Engineer logisticians must constantly track the status of orders throughout the requisition process to ensure that they are filled.

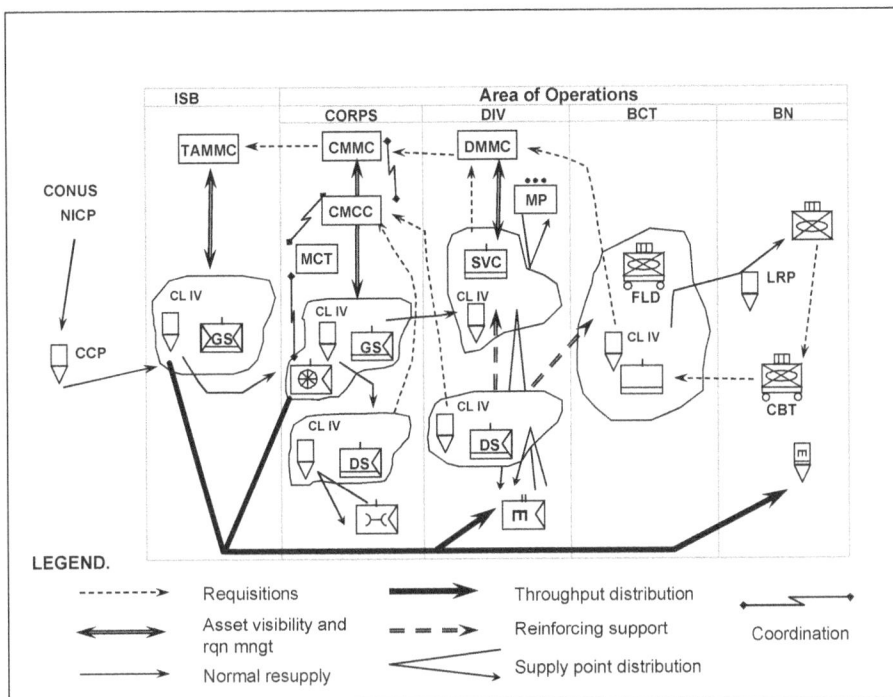

Figure 10-12. Class IV requests and distribution in contiguous AOs

10-15. Maintaining Class IV supply points is clearly a logistics function that engineer units are not organized or equipped for. Although engineer units should avoid responsibility for operating Class IV points, recent and repeated experience in contingency environments has shown that engineers habitually are forced to do so to ensure the completion of GE missions, particularly when time constraints exist. Engineers should be involved, but they should not be required to run Class IV points. Table 10-1, pages 10-10 through 10-12, is an example of a list of supplies that units might maintain in an engineer Class IV point during a contingency. Note that it contains only very basic materials and supplies. Units may need to be creative in the way they obtain Class IV supplies. Using materials from base camps that are closing is an example.

10-16. Engineers may also procure construction materials in theater using local purchase procedures or contracting. In a contingency, engineer logisticians must rapidly learn the methods and rules for obtaining construction supplies through the appropriate system. To maximize its benefits, local procurement should occur as close as possible to the actual construction site to minimize transportation requirements. Engineers must learn specific procedures and rules for local purchase procedures and contracting. Some of the options include—

- **Government-wide commercial purchase cards.** These are useful instruments for purchase of supplies up to an established limit. It is an effective method for small purchases. When deploying, users must determine the specific rules for their cards for the specific contingency. Depending on the deployment location, there may be problems with finding vendors who are willing to accept the government-wide commercial purchase cards.
- **Blanket purchase agreements (BPAs).** A blanket purchase agreement is a simplified method of filling anticipated repetitive needs for supplies by establishing charge accounts with qualified sources of supply. A blanket purchase agreement is a written understanding between the government and a supplier that eliminates the need for individual purchase and payment documents.

- **Prime vendor program.** This is a DOD institutionalized program that is operated by the Defense Logistics Agency. It establishes a series of contracts with different vendors. When a specific item is needed, each vendor is given an opportunity to bid to fill the order in a set period of time.
- **Logistics civilian augmentation program (Army).** The logistics civilian augmentation program (Army) is a program for preplanned use of a civilian contractor to augment capabilities of selected forces during a contingency. Units may obtain logistics support, to include Class IV through this program.

10-17.   Although obtaining materials for GE missions is often the most advantageous method for needed requirements, engineers must keep certain factors in mind while doing so. Some of these factors include—

- Standard sizes of GE materials may be different in the AO. Dimensional lumber is often cut to different standards in foreign countries. Voltage systems in overseas locations are also typically different from CONUS.
- The quality of different items may be considered substandard. Lumber, concrete, and asphalt are three examples of construction materials that are typically not consistent with U.S. standards.
- Language and cultural difference may make it difficult to obtain GE supplies. In some situations, local vendors may feel it is more important to try to please you in initial discussions than tell you the truth about whether they are capable of providing materials in the quantity and quality needed.
- Military operations may drive up prices. Shortages caused by multiple units competing for the same resource may induce local suppliers to profiteer from ongoing operations.

Table 10-1. Sample stockage level for engineer class IV point

Line		Nomenclature	NSN	Unit of Issue
4	AA	Sandbags	8105-00-142-9345	HD
4	AB	Wire, barbed	5660-00-224-8663	RO
4	AC	Wire, concertina	5660-00-921-5516	RO
4	AD	Pickets, long, 6 feet long	5660-00-270-1510	EA
4	AF	Pickets, short, 3 feet long	5660-00-270-1589	EA
4	AG	Barrier, Hesco bastion, 2 x 2 x 10 feet	2590-99-169-0183	EA
4	AH	Barrier, Hesco bastion, 2 x 2 x 4 feet	2590-99-001-9392	EA
4	AI	Barrier, Hesco bastion, 3 x 3 x 2.5 feet	2590-99-001-9393	EA
4	AJ	Barrier, Hesco bastion, 3 x 5 x 2.5 feet	2590-99-001-9395	EA
4	AK	Barrier, Hesco bastion, 4.5 x 3.5 x 2.5 feet	2590-99-835-7866	EA
4	AL	Barrier, Hesco bastion, 4.5 x 4 x 2.5 feet	2590-99-391-0852	EA
4	AM	Barrier, Hesco bastion, 7 x 7 x 7.5 feet	2590-99-335-4902	EA
4	AN	Lumber, 1 inch x 6 inches x 12 feet	5510-00-220-6080	EA
4	AO	Lumber, 1 inch x 4 inches x 12 feet	5510-00-220-6078	EA
4	AP	Lumber, 1 inch x 10 inches x 12 feet	5510-00-220-6084	EA
4	AQ	Lumber, 2 inches x 4 inches x 8 feet	5510-00-220-6194	EA
4	AR	Lumber, 2 inches x 4 inches x 10 feet	5510-00-220-6194	EA
4	AS	Lumber, 2 inches x 4 inches x 12 feet	5510-00-220-6194	EA
4	AT	Lumber, 2 inches x 6 inches x 8 feet	5510-00-220-6196	EA

Table 10-1. Sample stockage level for engineer class IV point (continued)

Line		Nomenclature	NSN	Unit of Issue
4	AU	Lumber, 2 inches x 6 inches x 10 feet	5510-00-220-6196	EA
4	AV	Lumber, 2 inches x 8 inches x 14 feet	5510-00-220-6198	EA
4	AW	Lumber, 2 inches x 10 inches x 12 feet	5510-00-220-6200	EA
4	AX	Lumber, 2 inches x 12 inches x 12 feet	5510-00-220-6202	EA
4	AY	Lumber, 4 inches x 4 inches x 8 feet	5510-00-220-6178	EA
4	AZ	Lumber, 4 inches x 4 inches x 10 feet	5510-00-220-6178	EA
4	BA	Lumber, 4 inches x 4 inches x 16 feet	5510-00-220-6178	EA
4	BB	Timber, 6 inches x 6 inches x 8 feet	5510-00-550-6825	EA
4	BC	Timber, 6 inches x 6 inches x 10 feet	5510-00-550-6825	EA
4	BD	Plywood, 1/2-inch x 4-foot x 8-foot ply	5530-00-128-5143	EA
4	BE	Plywood, 5/8-inch x 4-foot x 8-foot ply	5530-00-128-5147	EA
4	BF	Plywood, 3/4-inch x 4-foot x 8 foot ply	5530-00-128-5151	EA
4	BG	Nail, common wire, steel 5d	5315-00-010-4656	LB
4	BH	Nail, common wire, steel 8d	5315-00-010-4659	LB
4	BI	Nail, common, 3 inch 10d	5315-00-753-3883	LB
4	BJ	Nail, common, 3 1/4 inch 12d	5315-00-753-3884	LB
4	BK	Nail, common, 3 1/2 inch 16d	5315-00-753-3885	LB
4	BL	Nail, common, 20d	5315-00-753-3886	LB
4	BM	Screening, insect, nonmetal, 48 inches wide	8305-00-559-5047	YD
4	BN	Bolt, machine, 3/4 inch x 12 inches with nut	5306-00-550-3697	EA
4	BO	Washer, flat cadmium steel, 13/16-inch inside diameter, 2-inch outside diameter	5310-00-236-6478	EA
4	BP	Hinge, butt, steel leaves, 3 1/2 x 1 3/4 inch	5340-00-243-6193	EA
4	BQ	Hook and eye, door steel, 3 inches	5340-00-243-3224	EA
4	BR	Nipple pipe, steel galvanized, 1/2 x 4 inches long	4730-00-196-1547	EA
4	BS	Union pipe, galvanized, 1/2-inch pipe	4730-00-240-1674	EA
4	BT	Elbow pipe, galvanized, 1/2 inch x 90° angle	4730-00-278-4773	EA
4	BU	Elbow pipe, galvanized, 3/4 inch x 90° angle	4730-00-249-1478	EA
4	BV	Reducer, pipe, galvanized, 3/4 inch to 1/2 inch	4730-00-231-5650	EA
4	BW	Valve gate, bronze screw, 3/4 inch, class 125	4820-00-288-7567	EA
4	BX	Pipe, steel galvanized, 3/4 inch x 21 feet (threads)	4710-00-162-1019	EA
4	BY	Nipple pipe, steel galvanized, 3/4 x 4 inch long	4730-00-196-1500	EA
4	BZ	Nipple pipe, steel galvanized, 3/4 x 2 inch long	4730-00-196-1505	EA
4	CA	Union pipe, galvanized, female 3/4 inch, 300 pounds per square inch with water on gas	4730-00-240-1675	EA
4	CB	Coupling pipe, mall iron, 1/2-inch standard weight	4730-00-187-7612	EA
4	CC	Coupling pipe, mall iron, 3/4-inch standard weight	4730-00-187-7613	EA
4	CD	Cap pipe, galvanized mall iron, 1/2 inch	4730-00-231-2424	EA
4	CE	Cap pipe, galvanized mall iron, 3/4 inch	4730-00-231-2425	EA
4	CF	Primer adhesive for PVC pipe	8040-01-001-2705	PT

Table 10-1. Sample stockage level for engineer class IV point (continued)

Line		Nomenclature	NSN	Unit of Issue
4	CG	Pipe, PVC drain waste vent, schedule 40, 20 foot long, 2 inch diameter	4710-00-476-5870	EA
4	CI	Outlet box, 4 x 4 1/2 to 3/4-inch knockout	5975-00-159-0969	EA
4	CJ	Cover junction box, 4 inches square flat	5975-00-281-0057	EA
4	CK	Junction box, rectangular, surface mounted for switch or receptacle	5975-00-281-0090	EA
4	CL	Wire, electrical 6 round 20° Celsius, .162 inches nominal	6145-00-299-4456	EA
4	CM	Wire, electrical 6 round 60° Celsius, .449 inches nominal	6145-00-519-1332	FT
4	CN	Cable, power, electrical, 4 oval, 60° Celsius	6145-00-519-2718	FT
4	CO	Wire, electrical 6 round 75° Celsius, .481 inches nominal	6145-00-939-4951	FT
4	CP	Wire, electrical 6 round 90° Celsius, .19 single conductor	6145-01-204-6473	FT
4	CQ	Wire, electrical 6 round 90° Celsius, .13 single conductor, (white)	6145-01-204-6477	FT
4	CR	Wire, electrical 6 round 90° Celsius, .13 single conductor (blue)	6145-01-204-6478	FT
4	CS	Wire, electrical 6 round 90° Celsius, .13 single conductor (red)	6240-00-152-2987	EA
4	CT	Lamp, fluorescent, 48 inches nominal, 3,000 lumens	6240-00-990-8191	EA
4	CU	Lamp incandescent, 5.25 inches maximum, 1,650 lumens nominal	6210-00-865-8451	EA
4	CV	Fixture, lighting, 36 inches, translucent, diffuser	6210-01-395-9544	EA
4	CW	Cement, Portland, 94-pound bag	5610-00-250-4676	BG

PRODUCTION OF CONSTRUCTION MATERIALS

10-18.   Certain types of materials are typically needed in large quantities and are of great weight that engineers must produce them locally (or contract a supplier). Soil for fill, sand, and gravel are examples of materials typically obtained from local sources. To produce more refined products, engineers may need to further process materials to obtain construction materials, such as crushed rock, asphalt, and concrete. There are specialized engineer platoons, teams, and sections that handle production missions for most types of construction materials. Significant environmental considerations may be placed upon U.S. forces when creating or operating these sites. At a minimum, each of these operations has inherent significant FHP issues associated with them.

BORROW PITS AND QUARRIES

10-19.   Specific classifications of pits and quarries are shown in table 10-2. Borrow pits are the preferred source of construction aggregate and fill material when resources are scarce and material quality is not critical. They are similar to quarries except they tend to be smaller and generally require no blasting and minimal mechanical efforts. Materials in borrows seldom need to be blasted, crushed, or screened. Though the gravel, sand, and fines obtained in a borrow pit may not be as good as crushed stone, it is often acceptable. Equipment needed for a borrow pit includes dozers for grubbing and clearing, dump trucks for hauling, and scoop loaders, scrapers, or cranes with a shovel and dragline for loading.

Table 10-2. Pit and quarry classifications

Type	Material	Primary Use	Operation
Borrow pit	Soil, sand, and gravel	Subgrades, base course, and fill	Medium and light mechanical
Gravel pit	Gravel, coarse sand, and clay	Base course, surfacing, and fill	Medium and light mechanical
Alluvial pit	Clean gravel and sand	Aggregate for concrete and mixes	Heavy mechanical crusher and screen and wash
Dump pit	Mine spoil, slag, and overburden	Recycling and surfacing and aggregate	Heavy mechanical crusher and screen and wash
Hard rock quarry	Aggregate	Base course, surfacing, and aggregate for concrete mixes or concrete and asphalt mixes	Heavy mechanical crusher, screen and wash, and drill and blasting
Medium rock quarry	Aggregate	Base course, surfacing, and fill	Heavy mechanical crusher, screen and wash, and drill and blasting
Soft rock quarry	Cement material	Base course, surfacing, and roads and airfields	Medium and light mechanical

10-20. Borrow pits are best located at the tops of hills close to or on the construction site for ease of material handling. If borrow pits are located away from the construction site, careful consideration should be made in locating them to ensure efficiency of operation and causing minimal environmental damage and impact on the local population. In planning the GE mission, units should take into account the time required to close the borrow pits in the overall timeline.

10-21. Specific information for opening and operating a borrow pit is contained in chapter 4 of FM 3-34.465. Units may also use FM 3-34.468 in these operations. In the future, these manuals will be combined to form one multi-Service manual that covers all aspects of pits and quarry operations.

10-22. Quarries are similar to pits except that they generally require drilling, blasting, or the mechanical removal of aggregate to obtain suitable material for a GE mission. Although not specifically part of the quarry operation, planners may find it advantageous to collocate rock-crushing capabilities, asphalt plants, and concrete production facilities. Specific information on quarries is contained in FM 3-34.465.

10-23. In a contingency operation, if it is determined that a quarry is required to support GE efforts, extensive planning must occur to ensure the operation is efficient, meets production requirements, and conforms to applicable environmental considerations. Unless there is an extremely large construction project, it is likely that one quarry will support multiple GE missions, so determining its location must be considered in a holistic manner. Layout of the site consists of preplanning the location, dimensions, and arrangement of the quarry and the supporting roads and facilities. Planners must consider the mission, source geology, amount of overburden, equipment available, access, drainage, and traffic flow when locating a quarry.

CRUSHED ROCK PRODUCTION

10-24. Rock of specific size and gradation is needed for asphalt and concrete production. Crushed rock is used as the base course for roads and airfields. Rock from quarry operations and borrow pit material must be crushed, screened, and perhaps washed to meet standards for a particular design. See FM 5-472/NAVFAC MO 330/AFJMAN 32-1221(I) for more information on methods of testing materials for proper design characteristics. Also, FM 3-34.465 contains extensive information on aggregate production through rock-crushing operations. Almost all contingency operations will require some level of crushed rock supply, and units with a rock-crushing capability are only in the reserve components. Planners must be aware that moving and establishing a rock-crushing capability is a time- and labor-intensive operation that must be well planned to meet specific project time constraints.

10-25.   The rock-crushing plant must be sited within a short distance of the quarry and collocation of these operations may be ideal. It should be located on level ground with good drainage with adequate space for equipment, stockpiles, and maintenance areas. An adequate supply of water must be available for the washing process. This water may require a settling basin or some other method to mitigate the environmental impacts of the operation.

10-26.   The two most common rock-processing units have either a 75- or a 225-ton production capability per hour. Each plant consists of several large pieces of towed equipment, to include the crushers, screening equipment, washing equipment, and conveyers. The mobile crushing, screening, and washing plant is diesel-and electric motor-driven, and it consists of nine major components capable of producing a minimum of 150 tons per hour of aggregate suitable for cement or asphalt concrete. The components and accessories include the following:

- Primary jaw crushing unit.
- Secondary cone crushing unit.
- Surge bin unit.
- Tertiary cone crushing unit.
- Washing and screening unit.
- Dolly unit.
- Three power generators.
- Ten product conveyors.
- Water-pumping unit.

10-27.   All units are semitrailer and trailer-mounted and can be operated independent, tandem, or combined to meet aggregate production requirements. Planners must be aware that the actual output from a given plant differs from its normal capacity in that it is dependent on the specific product input, the desired size of the final product, and the proportion of the by-product.

10-28.   Maintenance of rock crushing equipment is a time-consuming process. Heavy loads and the abrasive action of the crushing operation, along with movement of large quantities of material, lead to wear and damage of the equipment. Repairs of older plants can be difficult because of a lack of spare parts. Dust, noise, and other environmental considerations must be taken into account when planning for the operation of a rock-crushing plant.

ASPHALT PRODUCTION

10-29.   FM 5-436 provides the doctrinal foundation for conducting asphalt production operations. Engineer units with organic asphalt production capability are low density and only reside in the Reserve Components. Projects that require asphalt must consider the complex process of moving and establishing an asphalt plant due to the lead time required. An adequate supply of raw materials, such as rock, sand, and bitumen must be on hand to conduct asphalt operations production. Asphalt equipment provides the asphalt team with the capability to produce large quantities (2,250 tons) of asphalt per day. The equipment can supply patch material for the maintenance of existing roads and highways; pave parking and storage areas, roads (3 to 4 miles per day), and airfields; and treat surfaces for dust suppression and stabilization. The asphalt equipment can be used through the full spectrum of conflict, but is most frequently used in stability operations. In offensive and defensive operations conducted in a contiguous fashion, it will typically be employed only in the rear area. The asphalt team is normally combined with other units (concrete, vertical, horizontal, and haul) to accomplish its mission. The asphalt team depends on the quarry and equipment support platoons to move the equipment and asphalt.

10-30.   The asphalt plant is a portable drum-type, electric motor-driven facility capable of self-erection (major components) and satisfactory operation without permanent-type footings. It consists of major units, components, and accessories as required to assemble a complete plant capable of producing 150 tons per hour of graded asphalt paving mix. The asphalt plant is trailer mounted and can be interconnected mechanically and electrically and operated to the rated capacity. A good road network is needed to avoid traffic jams and resultant cooling of mixes. The planner must also consider the potential environmental problems, including dust generated by the plant and potential soil contamination from bitumen and fuel spills.

10-31. The bituminous-material paving machine is a self-propelled, crawler-mounted, diesel-engine-driven machine with an 8-foot basic paving width. The paving machine is capable of laying, compacting, and finishing bituminous concrete strips 6 to 16 feet wide. The paving machine consists of a receiving hopper, a spreader, a compaction unit, cut-off shoes, and a screed with the capability of being extended.

10-32. The asphalt melter is a skid-mounted, 750-gallons-per-hour de-drumming asphalt melter. The de-drumming tunnel is capable of removing 85- to 100-penetration cement from twelve 55-gallon drums at one time. The unit also contains a 3,000-gallon hot-storage compartment for heating the asphalt to pumping temperature (235°F). The melter can operate individually, in pairs, or in trios in parallel from a single source of hot oil.

10-33. The hot-oil heater is a trailer-mounted, heavy-duty, high-output-capacity unit designed to transfer oil and pump it through transmission lines to the asphalt melter and storage tank. Fuel and external electric power are required for operation.

CONCRETE PRODUCTION

10-34. The 16S concrete mixer meets small-scale concrete requirements. The 16S mixer is ideal for small missions and can be moved to remote locations. It is manpower intensive, but can be grouped together to form a more efficient concrete mix operation.

10-35. The M5 engineer mission module concrete mobile mixer is transported by an M1075 palletized load system and an M1076 palletized load system trailer. The M919 concrete mobile is a self-contained concrete material transporter and mixing machine. It is capable of producing high-quality, fresh concrete at the construction site. The machine has the capacity to carry the materials for 5 to 8 cubic yards of concrete depending on usage (mobile/stationary). The M919 has limited trafficability and must remain on firm ground. It requires a scoop loader to support it while mixing.

10-36. FM 5-428 provides planners, designers, and general engineers using concrete in their construction with information on the production of this construction material. Planners refer to it when determining the design mixture required for a specific mission.

LOGGING AND SAWMILL OPERATIONS

10-37. The Engineer Regiment no longer maintains an organic capability to conduct logging operations and supply timber products for construction. These engineer forestry teams were divided into a logging and sawmill section. However, since these units are no longer in existence, planners must procure these products instead of producing them, or contract for HN or civilian teams to directly support engineer requirements with logging and sawmill operations if the demand is high enough. Some allies may have organic units in their military forces to conduct these operations.

This page intentionally left blank.

Chapter 11

Base Camps and Force Bed-Down Facilities

If you can figure out the criteria for base camp selection...you've done something the Army can use.

LTG Robert B. Flowers, 50th Chief of Engineers

For over two centuries, the Army and its sister Services (with the assistance of HN and contract support) have employed base camps as support locations for forward-deployed forces. Base camp construction is not a new idea. What is new is the recent trend to outsource base camp operations (food, power, waste, bed-down, and construction). Although advantageous under certain circumstances, outsourcing has proven expensive and sometimes wasteful. Outsourcing supplies and services is often a result of the political need to minimize the number of U.S. Soldiers deployed to an AO, resulting in base camps being very expensive to operate and maintain. Recent experiences in such places as Bosnia, Kosovo, Kuwait, and Iraq have become test beds for engineering and base camp construction. The quality of life for deployed Soldiers have become increasingly important, and base camp facilities have had to improve beyond the standards initially planned for by military leaders.

RESPONSIBILITIES

11-1. A base camp is an evolving military facility that supports the military operations of a deployed unit and provides the necessary support and services for sustained operations. It is a grouping of facilities colocated within a contiguous area of land, or within close proximity to each other, for the purpose of supporting an assigned mission, be it tactical, operational, or logistical. A base camp may be located near a strategic piece of real estate, such as a port, an airfield, a railroad, or another major LOC. Base camps support the tenants and their equipment; and while they are not installations, they have many of the same standards the longer they are in existence. Figure 11-1, page 11-2, shows the conditions of Camp Bondsteel, Kosovo, in July 1999 before construction. Figure 11-2, page 11-2, shows the conditions of the same camp in October 1999 after construction.

11-2. The primary purpose for a base camp is mission support—the support that a base camp system provides for the execution of the overall military mission of the deployed force. To execute mission support, a base camp must provide survivability and other aspects of protection to deployed forces, resource management of critical infrastructure, training opportunities for deployed forces and permanent party, and maintenance to facilities. Included in that mission support is the application of environmental considerations, to include the critical aspects of FHP.

COMBATANT COMMANDER

11-3. The CCDR is responsible for the major decisions involving base camp location and development within the AO. The CCDR may delegate authority for base camp decision making to component commanders or to commanders exercising 10 USC service responsibilities (the ASCC in the case of the Army). Decisions are often made in consultation with the HN, subordinate commanders, and U.S. Department of State (DOS) representatives. CCDRs provide the necessary planning and construction guidance for base camps in the form of a standards book or through OPLANs and OPORDs.

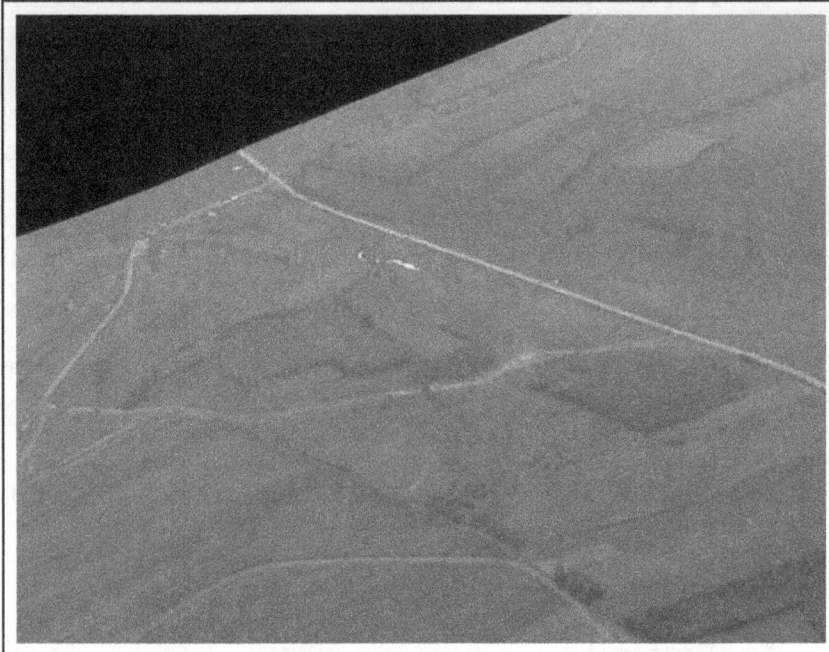

Figure 11-1. Camp Bondsteel, Kosovo, July 1999

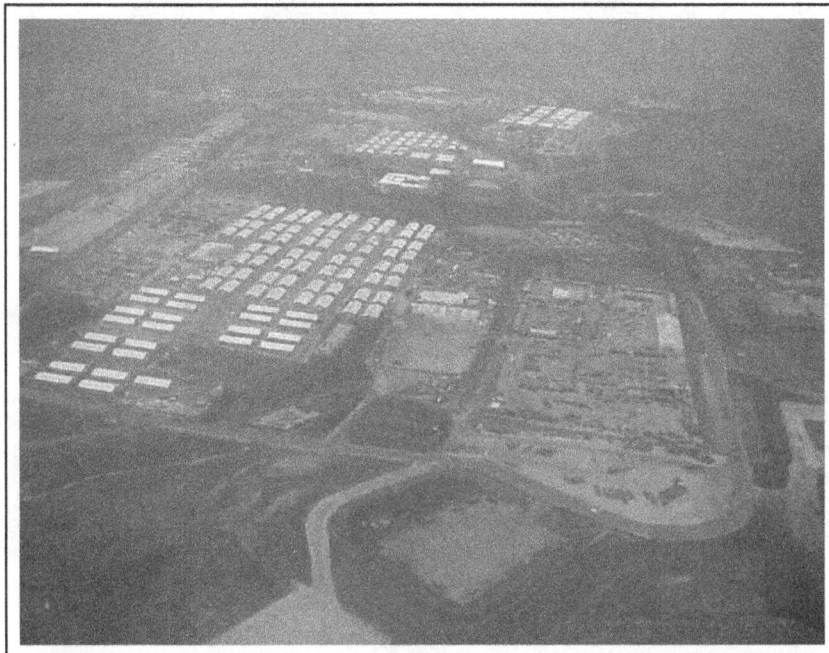

Figure 11-2. Camp Bondsteel, Kosovo, October 1999

ARMY SERVICE COMPONENT COMMANDER

11-4. The ASCC establishes a staff engineer section with a facilities and construction department that manages engineering and construction within the AO under the appropriate 10 USC responsibilities. This staff engineer section is responsible for developing the base camp and bed-down plan for all Service personnel and equipment arriving in the AOR. With guidance from the CCDR and the approval of the ASCC, they provide guidance on engineering and construction missions; establish standards for construction; conduct coordination with the HN; participate in funding, utilization, and resourcing boards; and coordinate with the USACE and the ENCOM. Their responsibilities include integrating the legal, FHP, and other aspects of environmental considerations provided from the respective areas of staff expertise.

THEATER ENGINEER COMMAND

11-5. The theater engineer command executes the majority of the ASCC engineering missions when they are deployed to support an AO, with either military engineering units or civilian contractors.

FORWARD-DEPLOYMENT AND REACHBACK CAPABILITIES

11-6. The ASCC engineer staff and subordinate commands rely heavily on FFE, in the form of forward deployment and reachback, to accomplish base camp design, construction, and management functions. See appendix D for a discussion on the teams and resources available, along with their specific capabilities. Note that the facilities engineering team is ideally suited to serve as a department of public works or camp mayor's cell for a forward-deployed base camp in a contingency operation.

11-7. The deployment of U.S. military forces has become increasingly common over the past two decades. These deployment missions include the critical task to set up and occupy a base camp that projects, sustains, and protects the force. Without proper master planning, synchronization, and oversight, base camp construction will not meet the needs of the commander. A lack of understanding of base camps as complex systems may result in very inefficient and ineffective base camps. Failure to include this discussion early in the planning process can cause significant problems for the force.

FACTORS

11-8. There are many factors that contribute to the overwhelming challenges and inherent difficulty in establishing base camps. Changing missions, fluctuating populations, turbulent civil-military situations, and an unclear end state all greatly affect the size, orientation, and stance of base camps. Another factor that hinders efficient planning of base camps in a postconflict situation is that units often begin establishing their base camps by occupying existing facilities where they ceased forward movement. It could be several weeks before a decision is made to establish a formal base camp in that location and when engineer teams arrive to begin planning.

PERSPECTIVE

During Operation Iraqi Freedom, the deserts of Kuwait became home to Soldier base camps. They bore names like Camp Doha, Camp Arifjan, Camp Victory, Camp Udairi, Camp New York, Camp New Jersey, Truckville, and Camp Wolf. In the course of only a few months, barren deserts became home to tens of thousands of Soldiers. The combined forces land component command engineer staff oversaw the planning, design, and execution of millions of dollars of construction effort as the United States prepared for the ground attack into Iraq. After crossing the border into Iraq, U.S. forces fought their way north, stopping for short periods of time to rest, rearm, and refit before moving again. Locations where the 3^d Infantry Division (ID) and 1st Marine Expeditionary Force (MEF) stopped for any length of time became small FOBs. As these FOBs were improved, several became actual base camps. As the tactical and COP became clearer over the following months, combined JTF-7 (Iraq) had designated 13 enduring base camps within Iraq.

STANDARDS

11-9. The CCDR specifies the construction standards for facilities in the theater to minimize the engineer effort expended on any given facility, while assuring that the facilities are adequate for health, safety, and mission accomplishment. Typically, the CCDR will develop the base camp construction standards for use within the theater using the guidelines provided in JP 3-34. Figure 11-3 shows the bed-down and basing continuum [and highlights the need for early master planning efforts to help facilitate transition to more permanent facilities as the operation develops. While the timelines provide a standard framework, the situation may warrant deviations. The engineer must recommend the most feasible solutions to each requirement based on construction guidelines and other planning factors. Other standards documents that provide specific construction examples include the United States Army, European Command (USAREUR) Red Book, Base Camp Facility Standards, and USCENTCOM Sand Book. The commander may also establish standards in specific OPLANs, OPORDs, and directives. These standards are used as initial guides and planning tools. Standards may also provide priorities for construction within base camps. Planners must be very familiar with the appropriate standards to execute construction and maintenance activities in a decentralized manner.

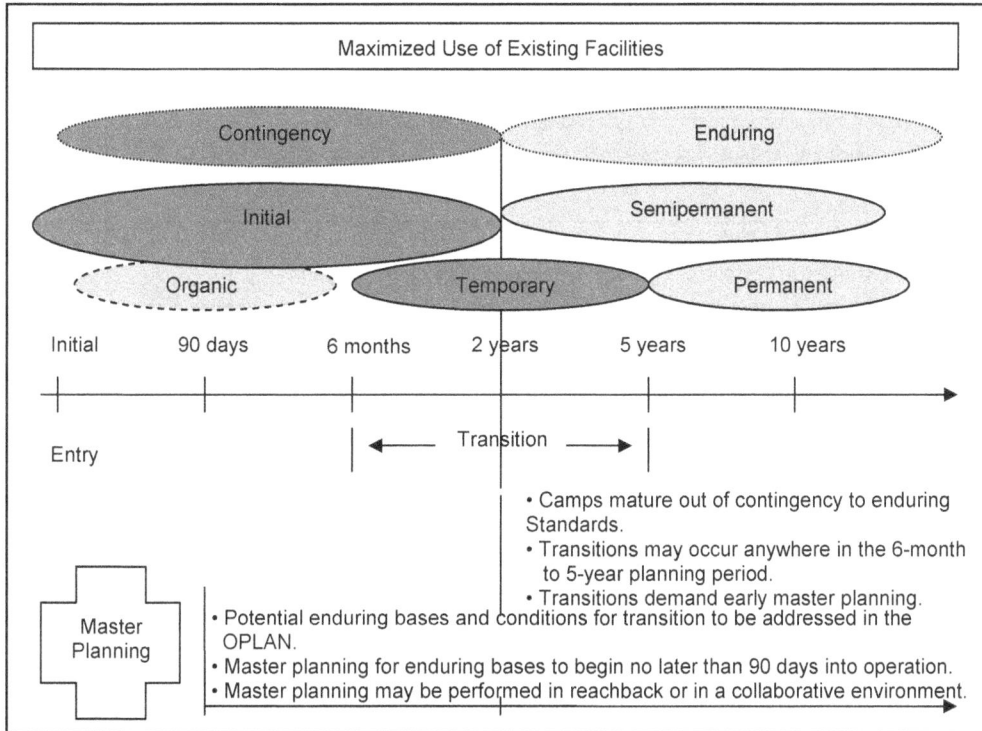

Figure 11-3. Force bed-down and base camp development

11-10.   There are two phases, based on anticipated lifespan, that identify the construction standards. Those phases are the contingency phase (0 to 2 years) and the enduring phase (over 2 years). Within each of the phases are subset standards which further refine the phases.

11-11.   Contingency phase standards (table 11-1, page 11-6) are defined by the following types:

- **Organic.** A subset of initial standard construction, organic standard construction is set up on an expedient basis with no external engineer support, using unit organic equipment and systems or HN resources. It is intended for use up to 90 days, and it may be used for up to 6 months. It typically provides for initial force presence and maneuver activities until force flow supports the arrival of engineer resources.

- **Initial.** Characterized by austere facilities requiring minimal engineer effort and the ease of material transportability or availability, initial standard construction is intended for immediate use by units upon arrival in theater for up to 6 months. Typical to transient mission activities, it may require system upgrades or replacement by more substantial or durable facilities during the course of operations.

- **Temporary.** Characterized by minimum facilities and effort with material transportability or availability, temporary standard construction is intended to increase the efficiency of operations for use extending to 24 months, but may fulfill enduring phase standards and extend to 5 years. It provides for sustained operations and may replace initial standard construction in some cases where mission requirements dictate and require replacement during the course of extended operations. Temporary standard construction can be used from the start of an operation if directed by a CCDR. It is typical to nontransient mission activities.

Table 11-1. Contingency construction standards in theater

Organic construction standards—
- Support on an expedient basis with no external engineer support.
- Use unit organic equipment and systems and/or HN resources.
- The mission duration is typically 1 to 90 days.
- Provide for initial force presence and maneuver activities until force flow supports arrival of engineer resources.

Initial construction standards—
- Are characterized by austere facilities requiring minimal engineer effort.
- Are intended for immediate operational use by units upon arrival for a limited time, ranging up to 6 months.
- May require replacement by more substantial or durable facilities during the course of operations.

Temporary construction standards—
- Are characterized by austere facilities requiring additional engineer effort above that required for initial construction standard facilities.
- Are intended to increase the efficiency of operations for use up to 24 months.
- Provide for sustained operations.
- Replace initial construction standards in some cases, where mission requirements dictate. The temporary construction standard may be used initially if so directed by the CCDR.

Type of Construction	Organic	Initial	Temporary
Site work	Minimal to no site work; maximized use of existing facilities	Clearing and grading for facilities (to include drainage, revetments, or POL ammunition storage, and aircraft parking) aggregate for heavily used hardstands; and soil stabilization	Engineered site preparation, including paved surfaces for vehicle traffic areas and aircraft parking, building foundations, and concrete floor slabs.
Troop housing	Unit tents	Tents (may have wood frames and flooring)	Wood frame structures, relocatable structures, and modular building systems
Electricity	Unit TACGENS	TACGENS: high- and low-voltage distribution	Nontactical or commercial power and high or low voltage
Water	Water points and bladders	Water point wells and/or potable water production and pressurized water distribution systems	Limited pressurized water distribution systems that support hospitals, dining halls, firefighting, and other large users
Cold storage	Contracted or unit purchased	Portable refrigeration with freezer units for medical, food, and maintenance storage	Refrigeration installed in temporary structures
Sanitation	Unit field sanitation kits and pit latrines	Organic equipment, evaporative ponds, pit or burnout latrines, lagoons for hospitals, and sewage lift stations	Waterborne to austere treatment facilities; priorities are hospitals, dining halls, bathhouses, decontamination sites, and other high-volume users
Airfield pavement[1]		Tactical surfacing, including matting, aggregate, soil stabilization, and concrete pads	Conventional pavements
Fuel storage	Bladders	Bladders	Bladders and steel tanks

1. The type of airfield surfacing to be used will be based on soil conditions and the expected weight and number of aircraft involved in operations.

11-12. DOD construction agents (USACE, NAVFAC, or other DOD-approved activities) are the principal organizations that design, award, and manage construction contracts in support of enduring facilities. Enduring phase facilities are categorized by the following two types:

- **Semipermanent.** This facility is designed and constructed with finishes, materials, and systems selected for moderate energy efficiency, maintenance, and life cycle cost. Semipermanent standard construction has a life expectancy of more than 2 years, but less than 10 years. The types of structures used will depend on the duration. It may be used initially if directed by the CCDR after carefully considering the political situation, cost, quality of life, and other criteria.
- **Permanent.** This facility is designed and constructed with finishes, materials, and systems selected for high-energy efficiency, low maintenance, and life cycle cost. Permanent standard construction has a life expectancy of more than 10 years. The CCDR must specifically approve permanent construction.

BASE CAMP LIFE CYCLE

11-13. When developing a base camp, there are required considerations and processes that contribute to the life cycle of a base camp. The life cycle is that time period from when the need is identified, continues through occupation, and is completed with transfer or closure. The factors that impact the life cycle (figure 11-4) are as follows:

- The planning and requirements.
- The design, construction, and camp operation.
- The transfer or disposal of real property at the end of the mission.

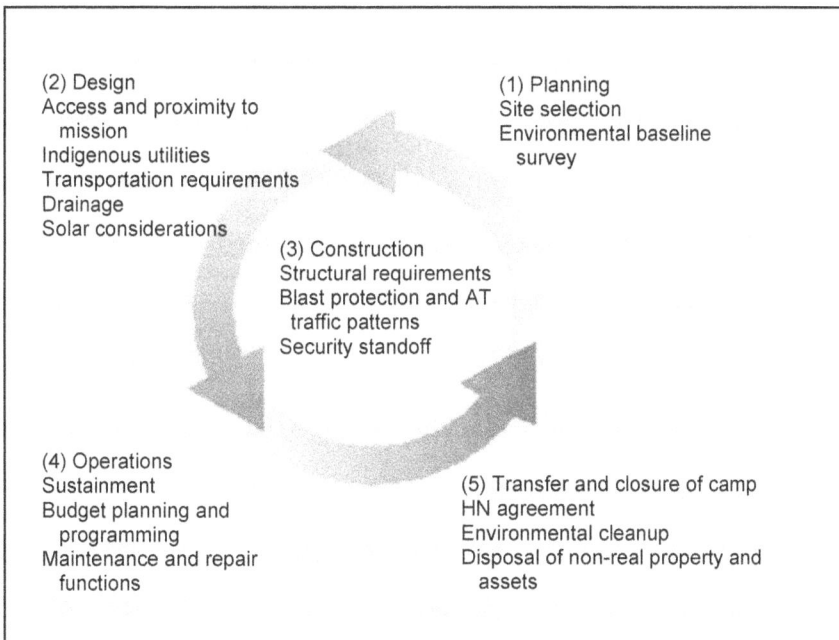

Figure 11-4. Base camp life cycle

BASE CAMP PLANNING

11-14. Engineers must be familiar with numerous planning considerations and design factors when planning the layout of a base camp. They must have and be familiar with the appropriate publications, references, and planning tools required to develop a base development site plan. A base camp development plan is the set of interrelated documents that record the planning process involved in laying out, determining the scope, and initiating implementation actions for a base camp during contingency operations.

11-15. There are four main components of the base camp development process (see appendix E). These components are listed below.

- **AT and protection.** AT and protection are critical to the development of contingency bases and long-term camps. Incorporating AT, physical security, and other protection concerns into site selection and the development of the base camp layout will ensure adequate protection of personnel and assets. The key to the effective development of base camp protection (to include AT, associated physical security requirements, and other protection considerations, to include survivability) is a partnership between those personnel focused on AT and other protection issues and the site engineers. This partnership helps to ensure the development of integrated physical security protective measures and security procedures that are consistent with base camp design. Essential to the base camp planning effort is the early identification of AT and other (to include physical security) protection requirements. Addressing the collective protection concerns early helps to ensure that site location and layout are compatible with security operations and mission accomplishment. See FM 5-103.
- **Facility standards identification.** The CCDR establishes the base camp standard for the joint operations area (JOA) by OPORD or FRAGO. These standards are intended to provide the CCDR's expectations to component commanders for base camp living and operating conditions.
- **Master planning.** Master planning provides an integrated strategy for construction and maintenance of required facilities at the best possible cost. The level of detail of the base camp master plan depends on the maturity of the location, the speed at which the operational need for a base camp develops, and the expected length of stay. A base camp master plan for expeditionary and/or initial standard camps may be simply a sketch of the camp, while a base camp master plan for temporary or enduring presence camps will include fully engineered construction plans based on complete surveys that integrate environmental considerations.
- **Construction management.** Responsible components (often USACE, theater command engineers, or facility engineer teams) will track the development of base camp construction according to the master plan priorities and report its progress.

11-16. Integral to this process is the expectation that the development of the base camp will have a limited timeframe and will require rapid planning and fast-track construction. Additionally, the possibility of rapidly changing military and political situations, a requirement to serve parallel missions in the same or neighboring regions, or a reintroduction of combat operations into the target area of the proposed base camp, may contribute to the need to alter the steps in the planning process. Also, the requirement to serve HN needs and concerns regarding the establishment of a single base camp or a series of base camps may change not only the described steps of the planning process, but also the options that may be considered with respect to the flexibility within each planning step. The intended life span of the facilities and infrastructure of a base camp depend upon mission-driven and economic decisions. A likely ingredient of this effort is the FFE support that the USACE will provide to the tactical commanders who determine the need for a base camp.

11-17. A decision to establish a base camp in a theater can be made at any time during the process of planning and executing a military operation. Ideally, it is made very early in the process to allow appropriate planning to take place in a proactive rather than reactive environment. The USACE usually has the mission to plan or assist in the planning and development of base camps in support of contingency operations. Engineers and planners must be prepared to support and assist users (whose first priority is the mission) in making base camp site selection and layout decisions. A base camp could be established in a hostile nation after active combat operations cease (such as in the case of Iraq), in a friendly nation as a contingency location to be used in the event of a deployment (such as in the case of Kuwait and Turkey), or in a friendly nation to support active combat operations in a nearby country (such as in the case of Qatar).

SPECIFIC TERMINOLOGY

Base Camp

11-18. A base camp is an evolving military facility that supports the military operations of a deployed unit. It provides the necessary support and services for sustained operations.

Base Camp Location Selection

11-19. Base camp location selection is a group of actions taken by a multidisciplinary team of U.S. personnel. Often base camp selection is done in cooperation with a HN, to locate and obtain land, water areas, transportation corridors, and associated air space to support a U.S. military mission.

Base Camp Development Plan

11-20. A base camp development plan is a time-sensitive, mission-driven, iterative, and cyclical process that determines and documents the physical layout of properly located, sized, and interrelated land areas, facilities, utilities, and other factors to achieve maximum mission effectiveness, maintainability, and expansion capability in the AO. Additionally, the process must address the eventual cleanup and closure of the base camp after the U.S. military mission is completed. Important products of base camp development planning include the planning report, maps, plan drawings, and other geophysical information. Further, the tabulation of existing and required facilities is essential in defining real-property assets, shortfalls and, subsequently, in developing projects and other actions to mitigate deficiencies. Some documents, such as selected maps, are used on a daily basis by assigned units and by those individuals who are responsible for operating and maintaining the base camp.

Base Development

11-21. Base development is the acquisition, development, expansion, improvement, construction and/or replacement of the facilities and resources of an area or location, either to support forces employed in military operations or deployed according to strategic plans.

Design Guide

11-22. A design guide is a written and graphically depicted set of standards that govern the design, development, visual aspects, and maintainability of a specific base camp. The guide may be used to define performance and customer service standards for various base operation functions.

Environmental Baseline Survey

11-23. An EBS documents the original environmental condition of the land. An EBS is required if an area is to be occupied by U.S. forces for more than 30 days. An EBS identifies environmental hazards and issues that could impact area suitability for occupation by U.S. forces. This document is also critical during base cleanup and closure, when the U.S. military prepares to return the land back to the HN in its original condition. For more information on conducting EBS before the establishment of a base camp, see the Environmental Baseline Survey Handbook: Contingency Operations (Overseas).

General Site Plan

11-24. General site planning includes the actions to inspect, analyze, and select suitable locations for the facilities, infrastructure, utilities, and other improvements to be located within the boundaries of a base camp. The result of this process establishes plan-view dimensions, corridors, zones, and boundaries for the development of a base camp, usually portrayed on overlays to maps of the area.

Land Use Plan

11-25. Land use planning is the process of mapping and planning the allocation of land use areas based on general use categories, mission analysis products, functional requirements and interrelationships, and criteria and guidelines. A land use plan is like a jigsaw puzzle because each piece of the plan is intended to fit together to form "a whole that is greater than the sum of the parts." The plan is sized and shaped to account for constraints that cannot be overcome, to take advantage of opportunities that exist, to accommodate existing requirements, and to allow for future expansion. Compatible land uses are placed close to each another, and incompatible land uses are not.

Site Design

11-26. Site design is sometimes referred to as "site planning" by design professionals. Site design includes the actions taken by a design professional to draw up and prepare detailed plans, specifications, and cost estimates for the construction or renovation of facility complexes, individual buildings, infrastructures, and supporting utilities. The term "site design" is used to avoid confusion with the terms "site planning" and "general site planning."

Base Camp Cleanup and Closure

11-27. Base camp cleanup and closure is the process of preparing and executing alternative COAs to vacate a base camp after a U.S. military mission is completed. An archival record is prepared that includes the operational history of the base camp and the actions taken to clean up and close the base camp, as well as a description of any cleanup and closure tasks that could not be completed that may lead to land use, health, safety, and environmental problems in the future.

Base Camp Development Planning Process

11-28. The base camp development planning process is depicted in figure 11-5. Planners rarely perform these steps in exact sequence; consequently, numbers are not assigned to the steps. At times, planners may enter the process when it is well underway. Planning is iterative and intuitive in nature. The base camp development planning requires a multidisciplinary, multistaff team approach to efficiently identify, analyze, and develop workable solutions to the many challenges that will require addressing. Base camp planning team members could include (most importantly) commanders and their staff from the units that will occupy or may already be occupying the base camp; operational planners and AT and protection experts; CA specialists; technical experts in engineering and other design professions; environmental and preventive medicine expertise; resource managers; range and training experts; program analysts; contracting, real estate, and other legal specialists; and HN planners.

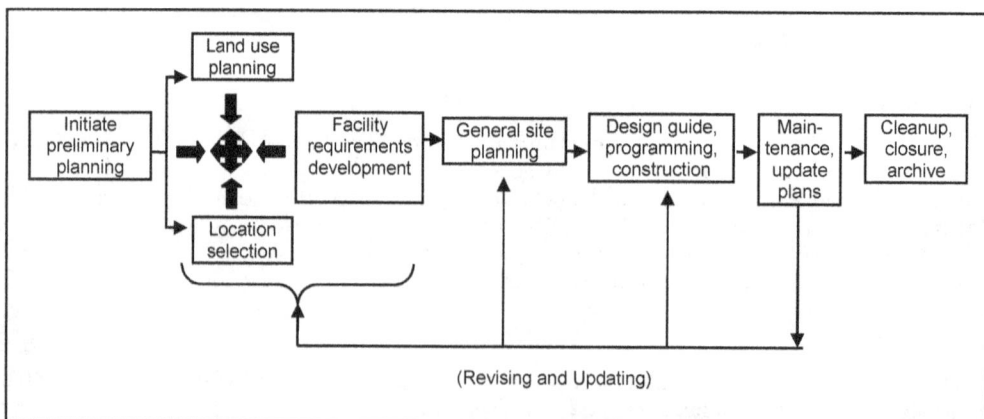

Figure 11-5. Base camp development planning process

11-29. Base camp development planning is an iterative process that is never finished until the facilities and land are turned back over to the HN. Base camp development plans must be shaped to always be reaching for improvements in the base camp's living and working conditions. For example, tents should give way to shelters. Shelters should give way to buildings. Field sanitation should give way to chemical, then waterborne, systems. Unimproved paths and roads should give way to pavements.

11-30. All levels of command are involved in real property planning and its related facility programming actions. Therefore, base camp development plans are reviewed and approved by not only the base camp commander or designated representative by means of a base camp planning board, but also by higher echelons as appropriate. This procedure has the added advantage of serving as a check-and-balance system against hasty or capricious planning. Additional technical review and approval of development plans for specialized projects and facilities, such as the planning of munitions storage and handling facilities, ranges and training areas, and high-security and aviation facilities, is required.

11-31. Planners should consider the objective end-state condition of the base camp facilities and the land area it occupies from the very start of the planning process. Initial agreements should address the cleanup, closure, and disposal or turnover of facilities that were occupied by U.S. forces to the HN. The objective condition of formerly occupied land must be thoroughly defined, because in many cases, the original owners want it returned in the same condition that it was in before U.S. occupancy.

11-32. The TCMS is the Army's official tool for base camp development planning and design. It is an automated military engineering construction planning and execution support system that delivers AFCS engineering and construction information for use in a TO. It provides military planners, logisticians, and engineers with the information necessary to plan, design, and manage theater construction projects where austere, temporary facilities are required.

DESIGN AND PLANNING CONSIDERATIONS

11-33. There is no single correct design to a base camp. There are an infinite number of designs, many of which are efficient and functional. Specific variables include whether the unit will be—
- Occupying existing facilities or building from the ground up.
- Using local labor and materials or bringing them in from other nations.
- Using a standards book or specific commander's guidance.

11-34. The design must also consider the operational aspects of the base camp and include a base camp zoning design and plan that will support the mission. Typical zoning considerations include the following:
- Zone 1, billeting.
- Zone 2, administrative areas.
- Zone 3, support facilities.
- Zone 4, storage facilities.
- Zone 5, motor pool and vehicle parking areas.
- Zone 6, waste and wastewater disposal.
- Zone 7, AT/protection measures.

Note. See the Base Camp Facilities Handbook for more information.

11-35. The base camp planning and estimating card in appendix E provides a quick planning method for the BCT or higher-level staff planners to develop an initial engineer estimate (such as area requirements, tonnage, man-hours, and equipment hours) for a base camp.

11-36. To determine the requirements for land, facilities, and infrastructure, Army base camp development planners must assess the mission, population, lifespan and construction standards, and commander's guidance. Each of these are discussed—
- **Mission.** Base camps are home to the force where the Army (and other Service elements) lives, works, trains, maintains, improves, and prepares for the next mission. The development of a base camp depends on the type of operations, missions, and activities anticipated for the encampment. Base camps may be classified as—

- **Main operations base camps.** Main operations base camps are continuously operated camps occupied by one or more large units with command, staff, logistical, and tactical functions. This base camp may be in support of a forward (C-130/C-17-capable) airfield and is primarily concerned with operational missions.
  - **Logistics hub base camps.** Logistics hub base camps (also referred to as logistics support areas) are continuously operated camps occupied by several large units with command, staff, and logistics functions. This base camp supports theater APOE, seaport of embarkation (SPOE), APOD, and SPOD and is primarily concerned with theater logistical support.
  - **Forward operating base camps.** FOBs are more remote, smaller, and austere camps that support operationally defined missions for a shorter period of time.
- **Population.** The population data of a base camp is fundamental information that is required for the base camp development plan process. As examples, the sizing of a base camp's utility systems and the determining of its billeting requirements cannot be accomplished without this information. Sources of population data for a base camp include—
  - Table of organization and equipment (TOE) documents.
  - Table of distribution and allowances (TDA) documents.
  - Nonappropriated fund and other U.S. government documents that provide data on other segments of the population, such as contractor personnel and local national employees.
  - The time-phased force and deployment data deployment list (TPFDL). *TPFDL* identifies types and/or actual units required to support the OPLAN, and indicates the origin and PODs or ocean area. It may also be generated as a computer listing from the time-phased force and deployment data.
  - The civilian tracking system provides information regarding U.S. civilians present, or scheduled to be present in the TO.
  - Population data for U.S. and foreign contractors, and HN employees can be acquired from the staffing numbers that accompany the U.S. government contracting documents that authorize contractors and local nationals. This personnel count must be added to the personnel count of the assigned military units to determine the total planned population of a proposed base camp.
- **Lifespan and construction standards.** The planned lifespan of base camps and facilities influences the standards used to design and construct them. Table 11-1, page 11-6, shows the joint contingency construction standards in theater based on anticipated lifespan as provided in JP 3-34.
- **Commander's guidance.** The CCDR's or ASCC's base camp facility standards give planning guidance and minimum construction standards. Further guidance is given in OPLANs, OPORDs, and FRAGOs. If a conflict arises between OPORDs, FRAGOs, and the facility standards book, orders take precedence. During the life cycle of a base camp, authorized facilities may progress from contingency to enduring or may be immediately established at any level depending on operational requirements and CCDR guidance. Meeting these standards may be a progressive effort.

11-37. The USCENTCOM Sand Book and USEUCOM Red Book provide very specific recommended minimum planning factors for the construction of facilities within contingency and enduring base camps. For bed-down facilities, the recommended minimum square footage for personnel accommodations is shown in Table 11-2. The table also shows how many personnel are housed in a Southeast Asia hut (SEAhut) or container. See Figure 11-6 for more information on SEAhuts.

Table 11-2. Recommended square footage for personnel accommodations

Category	Net Square Feet	Number Per SEAhut	Number Per Container (8 x 20)
E1 through E5	80	6	2
E6 through E7, WO-1/2, O1/2	130	4	2
E8, CW-3/4, O3/4	160	3	2
E9, CW5, O5/6	256	2	1
O7+	512	1	1

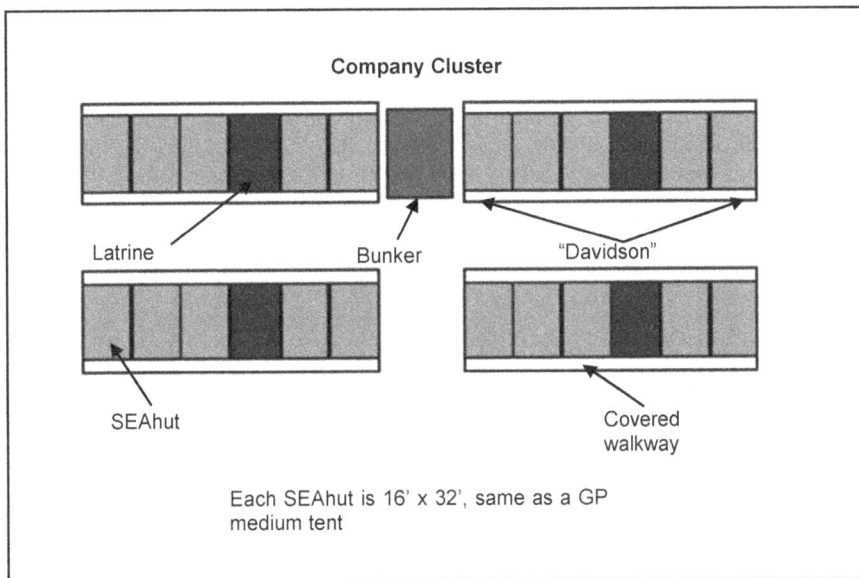

Figure 11-6. SEAhut company cluster

FIRE PROTECTION

11-38. Fire protection must be planned into the design of all base camps. Tent separations, wiring standards, and Soldier education are all critical components in reducing or preventing base camp fires and mitigating their effects on Soldiers and equipment. Over 50 tents in Kuwait were lost due to fires during Operation Enduring Freedom and Operation Iraqi Freedom. Most fires were due to improper electrical wiring connections and involved contractor-supplied tents that did not have the same flame-retardant material that the military-issued tents have. A lack of proper spacing, cleanliness, unit discipline, fire protection equipment, and training all contribute to this hazard.

Note. It is possible to retrofit tentage that is not flame-retardant.

UTILITIES

11-39. Utility system design must be based on current applicable TMs and guidance. Engineering calculations will be used to size the system. Where economically supportable and practicable, electric grids should be connected to commercial power. Smaller or remote bases should construct central power plants capable of supporting 125 percent of camp maximum demand load or use distributed generators of sufficient capacity to support maximum demand loads. When stand-alone, distributed generators are the main power source, they will be sized so that no generator set is loaded at less than 50 percent. The aspects of heating, ventilation, and air conditioning (HVAC) should be considered. All facilities where personnel are billeted or work should have heating by some means. Cooling in some climates is also more than a luxury item. For larger facilities, a central HVAC system will in all likelihood be more economical and practical when available. If it is economically feasible, tie into local municipalities and meet Army health and other protection standards. Installation of wells for potable water is authorized. There should be a minimum of two wells per camp—one primary and one backup (located within the camp boundaries). The last choice is to have potable water and/or bottled water trucked in. It also is more preferable to connect to a municipal sewage and wastewater treatment plant than to transport waste offsite for disposition.

LAND USE PLAN

11-40.   A land use plan depicts general locations for areas in relation to any existing development patterns and any existing major constraints that would have been identified by earlier data analysis. A land use plan should portray the basic scheme for main vehicular and rail networks, and it should designate the most advantageous locations and alignments for the mains, the stations, and the plants associated with the utility systems. Land use relationships should achieve the most efficient arrangement of functions, should resolve existing problems, and should provide logical and desirable locations for all mission and functional requirements. The distances shown in Table 11-3 should be used as minimum spacing for specific types of facilities. Drainage considerations must be applied (see appendix E).

Table 11-3. Minimum distances between facilities (in feet)

	Solid Waste	Ammunition	Helipad	Maintenance	Parking Lot	Roads	Billets	60kW TQG	Bulk Fuel	Potable Water	Wastewater	Shower	Laundry	Food Service	Latrine
Latrine	300	300	500	200	200	15	200	50	300	100	200	0	50	300	
Food Service	300	300	500	300	300	15	200	50	300	50	200	300	300		
Laundry	300	300	500	200	200	15	200	50	300	50	200	50			
Shower	300	300	500	200	200	15	200	50	300	50	300				
Wastewater	300	200	300	200	200	15	200	50	200	300					
Potable Water	300	200	300	300	300	15	200	50	200						
Bulk Fuel	300	500	300	200	200	15	200	200							
60kW TQG	50	300	300	50	200	15	50								
Billets	300	200	500	200	200	15									
Roads	15	15	300	15	15										
Parking Lot	20	20	30	50											

Table 11-3. Minimum distances between facilities (in feet)

	Solid Waste	Ammunition	Helipad	Maintenance	Parking Lot	Roads	Billets	60kW TQG	Bulk Fuel	Potable Water	Wastewater	Shower	Laundry	Food Service	Latrine
	0	0	0												
Maintenance	30 0	30 0	30 0												
Helipad	30 0	30 0													
Ammunition	30 0														
Solid Waste															

SPECIFIC FACILITIES WITHIN BASE CAMPS

11-41. Used extensively in recent contingencies, SEAhuts, as shown in a cluster configuration in figure 11-6, page 11-13, are 512 square feet (16 by 32 feet). A SEAhut has eight 110- or 220-volt electrical outlets. Normally, there is an environmental control unit (ECU) on each end for climate control. SEAhuts are constructed of wood with a sheet vinyl floor, 5/8-inch gypsum walls and ceiling, flat latex paint, metal roof, precast concrete pilings, painted exterior, and nail board 6 feet above the floor (so Soldiers can put a nail on the wall to hang things). When in a "Davidson" configuration, there are 5 SEAhut units, with a 12- by 32-foot latrine, for a total of 2,944 square feet of enclosed space. There is a 5-foot-wide walkway on each side. An administrative configuration has 3,072 square feet, but the latrines only take up 256 square feet. The building has walkways all around the building. The entire footprint is 42 by 106 feet, including walkways.

LIFE SUPPORT AREAS

11-42. A standard life support area has 20 temper tents in the configuration shown in figure 11-7. The large spacing between tents is for fire lanes and to allow cranes and other heavy equipment to move around to service air conditioners without damaging the existing tents or wires. The wide fire lanes provide a firebreak and maneuver room for firefighting equipment. Wooden buildups should not be constructed on or inside the tents, as the wind tends to drag the tent fabric across the rough wooden edges and destroy the tents. The tent city has a 4-inch gravel pad surrounded by a ditch. The ditch is for drainage when it rains and to separate the "no drive" area of the life support area from the rest of the camp. Its size and depth should be adequate to contain anticipated rainfall and runoff as well as prevent vehicular traffic from entering and exiting at nondesignated points across the ditch. Geotextile is normally not used.

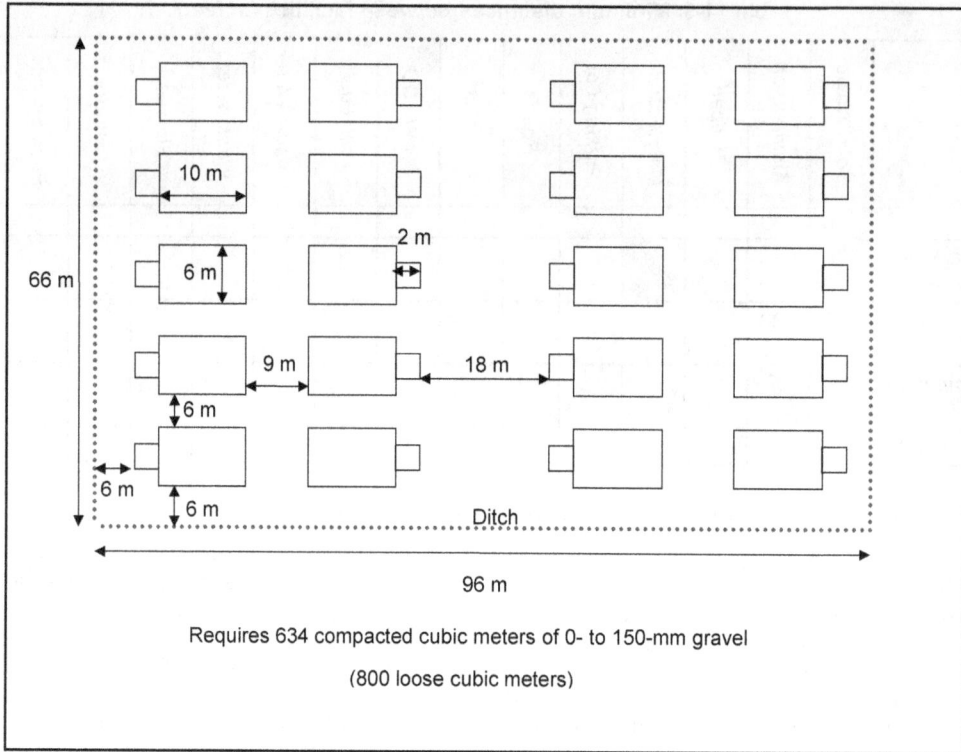

Figure 11-7. Standard life support area

SURGE HOUSING

11-43.   Base camps should maintain the ability to house 10 percent of the total population as transients and surges. During surge periods that exceed 10 percent, Tier 2 tents (maximum) will be used for housing. The definition of construction standards for tents include—

- **Tier 1.** Tier 1 consists of a GP medium field tent with plywood floor panels.
- **Tier 2.** Tier 2 consists of a GP medium field tent with plywood floor panels, two electric light outlets, two electrical outlets, and space heaters.
- **Tier 3.** Tier 3 consists of a GP medium field tent, full wooden frame for tent, plywood panel sidewalls, raised insulated flooring, four electric light outlets, eight electrical outlets, and space heaters.

TOILET AND SHOWER FACILITIES

11-44.   Toilet and shower facilities will be lighted, heated, and equipped with hot and cold water. Sanitary wallboard is the preferred wall covering for latrines. Sheetrock, if used, must be waterproof with a waterproof finish for cleaning. The female to male facility ratio will be based on the actual percentage of the sexes on a base camp at the current time or anticipated for the near future. The goals for all base camps are as follows:

- **Showers.** A shower head with a population ratio of 1:20 to 1:10.
- **Toilets.** A toilet with a population ratio of 1:20 to 1:10.

OTHER ADMINISTRATIVE AND SUPPORT FACILITIES CONSIDERATIONS

HEADQUARTERS FACILITIES

11-45. The size and type of headquarters facilities will depend on the situation and be based on the established standards. The task force headquarters facilities should include at least the following:

- Fencing.
- Parking.
- Security lighting.
- Secure compartmentalized information facility.
- Facility to house the tactical operations center operation.
- Commander's office space or building.
- Space for primary staff offices.
- Communications platforms and shelter.
- Command bunkers, guard shacks, and so forth.

11-46. Separate headquarters facilities should also be considered for brigades, battalions, and company elements.

LOGISTICAL SUPPORT FACILITIES

11-47. The construction of facilities such as a supply support activity, a direct exchange facility, a central issue facility, and warehouse facilities (when required), will depend on equipment density, troop strength, and classes of supply to be supported and will range from fixed warehouse facilities for long-term storage and the use of military vans (containers) for transportation and short-term storage. See chapter 12 for a more detailed discussion on support area facilities.

DINING FACILITIES

11-48. Dining facilities should have dining room space for their patrons as well as kitchen, administration, and storage space (open and refrigerated). Adequate space for cleaning, latrines, and clothes changing/locker areas for local national/contract kitchen staff may also be required.

FINANCE AND PERSONNEL SUPPORT

11-49. When finance center and personnel support operations are required, working areas and separated customer service areas should be included. For the finance detachment, this may include space for a pay cage and a vault.

POSTAL FACILITIES

11-50. When a postal facility is required, the building may not exceed the standards in DOD 4525.6-M, chapter 13. Postal facilities must also be at least 20 meters from any other structure.

LAUNDRY COLLECTION AND DISTRIBUTION POINT

11-51. Each base camp as it develops should have a laundry collection and distribution point to serve its population.

AVIATION FACILITIES

11-52. When aviation elements are present on a base camp, there are many aspects that must be taken into consideration to fully address their requirements. Aviation assets need things such as—

- Helicopter landing and parking pads.
- Vehicle parking areas.

- A lighted landing pad or runway.
- A forward area refuel point (FARP).
- A control tower.
- A squadron operations area.
- Areas for aviation maintenance and ground maintenance.
- Helicopter and aviation washracks.
- Other areas.

Note. Areas where uploaded ammunition is on aircraft "hot pads" or stored in close proximity should be separated to create a standoff distance or protected in a manner to minimize damage from an accidental weapons discharge. See chapter 6 for a more detailed discussion on aviation facilities.

COMMUNICATIONS FACILITIES

11-53.   When a communications compound or network service center is authorized, the facility size will depend on its purpose and the intended number of occupants it is to support.

MEDICAL FACILITIES

11-54.   The level of medical support and type of clinics and hospitals will vary, but should be taken into consideration when planning base camps. The specifics range from aid stations through clinics (dental and medical) to field hospitals. The actual requirements will directly relate to the mission, medical, and dental support requirements and the expectations of the command.

MOTOR POOL FACILITIES

11-55.   The need and space requirements for motor pool facilities and equipment parking areas must be recognized early on in the base camp planning and synchronized with the traffic flow patterns of the base camp and its security planning. Space for the conduct of maintenance (enclosed and exterior maintenance pads) and an area for administrative functions are requirements in almost all operations where base camps are established. This may also generate a need for DS maintenance facilities as well.

FUEL STORAGE FACILITIES

11-56.   As the length of the operation grows, it is more economical and safe to transition from using fuel bladders for fuel storage to aboveground fuel tanks. Keep in mind the environmental requirements and the need for secondary containment.

HAZARDOUS-WASTE COLLECTION POINTS

11-57.   The need to address hazardous wastes generated by our operations does not go away in a deployed environment (see the environmental annex of the applicable plan or operations order for specific requirements). HAZMAT spill kits should be kept on site and available.

PARKING LOTS

11-58.   In addition to motor pool equipment parking areas, depending on the size and purpose of each facility contained on a base camp, there may be a requirement to provide parking areas. These, as all developed parking areas, should be constructed using well-graded, compacted rock and soils with an engineered slope and drainage to minimize weather effects and increase the safety and longevity of the parking area.

KENNELS

11-59. A lighted and climate-controlled kennel facility and an exercise yard may be required when military working dogs are attached.

MORGUE

11-60. Considerations for a morgue facility include—
- Separation from the medical facilities.
- Privacy.
- The need for a work space.
- The requirement for refrigeration.

AMMUNITION SUPPLY POINTS

11-61. ASPs vary in size and purposes. They include the need to establish basic load ammunition holding areas and captured ammunition holding areas. Sometimes, these are colocated but segregated; other times, they are separated into completely different locations. Storage in bunkers is preferred, but storage in ammunition-certified containers located in bermed cells is another option. The least-preferred option is open-air storage whether in bermed cells or not. See chapter 12 for more detailed discussions on ASPs.

WASHRACKS

11-62. There is a need for washrack facilities in prolonged operations to support the maintenance of vehicles and the conduct of operations. Some considerations are that they should be designed to fit the largest and heaviest vehicles in the flee, and where possible, the washrack should be equipped with oil-water separators.

FIRE PROTECTION

11-63. Fire and emergency response services should be provided according to DOD guidance when practical and available. Fire stations should include accommodations for the firefighters, the fire trucks and associated equipment. There may be a need for a "live-fire" training area to keep the firefighting skills honed and the equipment exercised. Siting of the fire station should take into consideration the response times to mission-critical areas requiring protection (such as an airfield, helipad, or headquarters). Locations that do not meet the criteria for having dedicated fire and emergency response services will still run a unit fire prevention program and assign extinguisher duties to properly trained personnel. See AR 420-1, Department of Defense Instruction (DODI) 6055.6, and FM 5-415 for more details on these requirements and Army firefighting capabilities.

TRAINING FACILITIES

11-64. The quantity and types of training facilities will depend on the size and mission of the unit as well as the specific base camp population. When training facilities are made available, the options vary from virtual trainers ranges to maintain basic soldier weapon proficiencies, to full-blown, live-fire ranges for the conduct of combined arms training. See DA Pamphlet 350-38, chapter 16, for specific requirements for live-fire ranges.

CHAPELS

11-65. When authorized, chapel design should be nondenominational with office space provided for the chaplain. The chaplain's office space should be designed to support privileged communications with Soldiers.

EDUCATION CENTERS

11-66.   In enduring operations, an education center may be authorized.

ARMY AND AIR FORCE EXCHANGE SERVICE

11-67.   Army and Air Force Exchange Service (AAFES) facilities (such as a PX, food and service concessions stands, and a barber and beauty shop) may be authorized and can be colocated for efficiencies.

MORALE, WELFARE, AND RECREATION FACILITIES

11-68.   When authorized, morale, welfare, and recreation (MWR) facilities considerations include—

- Fitness facilities.
- Multipurpose facility (field house, theater, and so forth).
- Athletic fields.
- Running paths or tracks.
- Armed forces network facilities.

PROTECTION CONSIDERATIONS

11-69.   AT, physical security, survivability, and safety during deployed environments present unique challenges to planners, engineers, and security forces. As is the case for fixed permanent facilities, the type and severity of the threat as well as the desired level of protection may be the primary considerations in the selection of the measures used. These considerations will affect decisions on many varied issues, such as the types of vulnerability reduction measures, the physical layout of facilities, facility groupings, and the entire infrastructure of the base camp. Important factors in planning security measures include the availability of existing facilities, the types of structures in which people live and work, existing natural and man-made features, types and quantities of indigenous construction materials, available real estate, and layout of utilities and other base infrastructure. For preexisting buildings, the standards for existing buildings should be used. Other factors for consideration are issues like facility access, standoff or separation between facilities, threat specific standards, power generation and distribution, base camp perimeter (fences, gates, guard towers, constructed fighting positions, and so forth), clearing barrels, walkways, buried utilities, bunkers, and water and fuel storage protective measures.

OPERATIONS AND MAINTENANCE CONSIDERATIONS

11-70.   Major facilities and base camps constructed for use in contingency operations are expensive to build and maintain. The goal should be to maximize the life expectancy of these facilities with minimum cost. Considerations include paint, signage, routine upkeep, and preventive maintenance (electrical, plumbing, exterior, and interior), road and pavement repairs, erosion control, snow and ice clearance, and dust abatement. See chapter 13 for more details on specific maintenance considerations.

Chapter 12

Support Area Facilities

Simplicity is the ultimate sophistication.

Leonardo da Vinci

Adequate support area facilities are as vital for combat operations as they are critical to sustainment operations. Engineers at the strategic, operational, and tactical levels construct, maintain, and repair facilities for receiving, storing, and distributing all classes of supply and supporting all other logistics functions. This chapter addresses the procurement, construction, maintenance, and repair of logistics facilities and associated environmental considerations, both for general supply and for the more specialized purpose of storing munitions. Engineers tasked to support logistics installations have the following four major missions: provide new facilities, maintain existing facilities, recover and repair facilities damaged by hostile actions, and upgrade existing facilities to meet minimum standards or usage requirements. In some combatant command AOs, peacetime construction and HN agreements have provided extensive facilities. In less-developed theaters, there may be no preexisting logistics facilities. In such theaters, adapting and converting commercial property to military use or constructing new facilities may be required to provide logistics support facilities. Due to the magnitude of new construction and maintenance and repair of existing infrastructure generally associated with support facilities, it is recommended that planners consult the following publications for planning and design criteria:

- AR 415-16.
- TM 5-301-1.
- TM 5-301-2.
- TM 5-301-3.
- TM 5-301-4.
- TM 5-302-1.
- TM 5-302-2.
- TM 5-302-3.
- TM 5-302-4.
- TM 5-302-5
- TM 5-303.

SUPPLY AND MAINTENANCE FACILITIES

12-1. Logistics support areas in a contingency environment vary widely. The simplest installation may be a hardstand surface with rudimentary surface drainage and a supporting road system. More complex installations may look like urban industrial parks, including warehouses; maintenance and repair facilities; water, sewage, and electrical utilities; refrigeration or other climate control capabilities; and supporting roads, railroads, ports, airfields, protective fencing, fire services, and personnel support administration facilities. Logistics installations include general, ammunition, and maintenance depots; storage sites (to include fuel storage); and hospitals.

CONSTRUCTION RESPONSIBILITY

12-2. The CCDR, JFC, and ASCC with 10 USC responsibility, identify the minimum-essential engineering and construction requirements for facilities, including new construction and repair of war-damaged facilities. For the ASCC, the theater engineer command is normally responsible for planning, prioritizing, and tasking subordinate units for project execution. The theater engineer command can provide construction assistance and restoration support to the other services when assets are available or as directed by the ASCC. Support may also be provided to multinational forces when they are assisting U.S. operations. The CCDR or JFC may designate a regional wartime theater construction manager (TCM) to coordinate and prioritize engineer construction activities of all services in a geographic area. Detailed command and support relationships are given in FM 100-16.

PLANNING CONSIDERATIONS

12-3. It is necessary to determine requirements for time-phased facility construction, war damage repair, construction materiel, and other engineering needs for supporting deployed U.S. forces. In developing and evaluating alternatives, planning should result in—

- Determining critical requirements, the duration of construction projects, and information for scheduling and requisitioning.
- Developing a logical task sequence based on priorities necessary to accomplish the mission.
- Determining an accurate estimate of the required materials and labor that takes into account HN guidelines and resources.
- Determining command and support relationships and providing for engineering coordination throughout the theater or AO.
- Identifying a method of controlling the situation as it develops or changes.
- Identifying environmental considerations that may impact planning decisions.
- Meeting the commander's guidance if possible.

SITE SELECTION

12-4. Planners conduct a preliminary reconnaissance (sometimes a map reconnaissance) usually followed by a field reconnaissance. The field reconnaissance team may be composed of, but is not limited to, representatives of those units which the facility will support, the S-3 of the unit responsible for construction, a command group representative, a CA personnel representative (and other specialists, such as real estate, medical, chemical, legal, environmental, and others as required), and a representative of the HN. Emphasis should be placed on the following considerations:

- Tactical situation.
- Capability to defend the site.
- Terrain.
- Availability of suitable existing facilities that may be either occupied immediately or modified to desired specifications.
- Environmental restrictions that may limit the size of the required facility (these may be caused by weather or HN policy) or that may affect the location, design, or requirements of facilities.
- Accessibility to projected traffic.
- Availability of construction materials.
- Climatic extremes that may demand refrigeration or other climate control measures.
- Potential mission expansion and surge requirements.

PROTECTION

12-5. Protection of a facility or installation may be accomplished by active and passive security measures, including facility hardening and dispersion. The enemy situation must be evaluated as thoroughly as possible. Threats to supply and maintenance facilities may include conventional and nonconventional ground forces, CBRN threats, and attacks delivered by direct and indirect systems. Remote delivery of mines should also be considered. In an asymmetrical AO, insurgent activities may pose a threat to logistics assets. In determining how to best protect a facility against interdictory attacks, the commander must take into account the surrounding terrain, the weather, the availability of Class IV and V materials to support protective measures, and the enemy situation. Another consideration that may influence the commander's decision is the HN's policy governing construction and the use of construction resources. See chapter 9 for more information on protection. FHP and other associated environmental considerations should always be a factor in assessment and planning. For facility protection, planners should consider using facility hardening, dispersion, standoff, and security:

- **Hardening.** Hardening of facilities should be emphasized when terrain constricts dispersion and the threat analysis indicates that the facilities are possible targets for enemy weapons. Hardening techniques are discussed in FM 5-103.
- **Dispersion.** Where terrain conditions permit, facilities should be dispersed to prevent the enemy from inflicting massive damage in a single strike. Precautions must be made, however, to ensure that operations are not unduly hampered by ill-planned dispersion schemes.
- **Standoff.** One of the most important considerations in protecting facilities is standoff. Standoff is the distance between the facility and the location from which the facility could be attacked. Increased standoff distance degrades the effect of explosive devices, helping to minimize blast effects.
- **Security.** Generally, security includes active and passive measures taken to thwart enemy troop interdiction. Active measures may include the construction of fighting positions, wire obstacles, earthen and concrete barriers, minefields, placement of remote sensors, and use of security patrols. Passive measures may include the use of CCD discipline (see FM 20-3). Refer to AR 190-11 for the required security measures for ASPs. Engineer tasks that support security measures include clearing a right of way for security fences and constructing guard posts, fences, and lighting systems. Protective minefields and/or ECP and entry control facility (ECF) functional zones may be required in some cases.

LAYOUT

12-6. When locating and positioning a support area facility, the commander evaluates all information gathered in the planning and reconnaissance phases. Once the commander or designated representative has made a decision on where the installation is to be built, the engineer develops a construction plan that takes into consideration available resources (military, HN, or contract construction personnel, materials, and equipment). The layout should be well coordinated and organized in such a way that it can be completed soon enough to meet the operational priorities and minimize future controversies.

12-7. Internal operating efficiency must also be considered in the layout. The TCMS and the AFCS show typical installation layouts. New construction and the use of nonstandard designs must be held to a minimum. Whenever feasible, facility requirements must be met by existing facilities (U.S. and HN), organic unit shelters, and portable or relocatable facility substitutes. Standards for new construction (initial or temporary) are dictated by the CCDR or ASCC based on the expected duration of use, availability of materials, man-hours of construction effort, and material cost. Locally available materials may dictate design and construction criteria. Plans are provided for many supply and maintenance facilities in the TCMS and the references located in the chapter introduction. Modifications may be required.

CONVERSION OF EXISTING FACILITIES

12-8. During many operations, the use and modification of existing facilities is more advantageous than new construction, based on the availability of time, labor, and materials. Chapter 13 discusses the procedures for acquiring existing facilities and other real property. HN agreements may require compensation for using or converting such facilities. Army engineers, HN parties, and civilian contractors are encouraged to use ingenuity, imagination, and inventiveness to adapt existing facilities for military use. A cost benefit analysis will be one of the factors used to determine if new construction is more prudent or appropriate. An infrastructure reconnaissance (assessment or survey) is recommended to document the condition of, and preexisting deficiencies in, existing structures adapted for military use. An EBS and EHSA conducted in concert with each other are recommended to ensure that HAZMAT which would endanger Soldier health are not present in the existing structures or their surrounding areas and to limit claims against the government later in the life of the facility. If not able to be coordinated initially, the EBS and EHSA must be completed as soon as possible.

12-9. Units and organizations that occupy facilities are encouraged to establish internal teams that perform routine maintenance and repair of facilities. Army engineers perform maintenance and repair work that exceeds the capabilities of user units. This support usually requires specialized skills or heavy equipment. Further information on real property maintenance activities (RPMA) is given in chapter 13.

AMMUNITION STORAGE AND SUPPLY

12-10.   A mature theater requires a network of ammunition supply and storage facilities. Well-situated and stocked ammunition storage and supply facilities are critical to the timely distribution of the required munitions. Ammunition must be stored with maximum attention to protection against natural and man-made threats, including accidents caused by careless storage and handling. Class V and Class V (W) (aircraft ordnance) supply items are explosive and often contain sensitive components. Improper, careless, or rough storage and handling of ammunition and explosives may result in malfunctions and cause accidents that result in the loss of life, injury, or property damage. Properly designed, constructed, and maintained ammunition storage and supply facilities will help limit the possibility of such accidents. Appropriate storage ensures the maximum serviceability and shelf life of stocks and reduces maintenance requirements to a minimum. Planners must address concerns contained in DA Pamphlet 385-64 to ensure the effective and safe storage of ammunition.

THEATER STORAGE LOCATIONS

12-11.   The theater distribution element coordinates storage and distribution of theater munitions. Ammunition may be pushed forward to the ordnance ammunition units, where further distribution is made to forward ASPs. These are located in a secure area. Units may then draw directly from the ASPs. Ammunition may be brought further forward to ammunition transfer points (ATPs) where munitions are transferred from corps haul assets to user resupply vehicles. Generally, the farther to the rear the ammunition facility is, the more elaborate the construction, and the more extensive the construction support required.

CONSTRUCTION RESPONSIBILITIES

12-12.   Engineer units are charged with construction responsibilities in support of ammunition storage and supply operations. These responsibilities include the following:

- Reconnoitering and improving and/or constructing roads and bridges, which provide access to and egress from the ammunition facility. Engineers will also construct roads within the facility.
- Locating water sources for firefighting operations and constructing required reservoirs and water distribution systems.
- Constructing standard ammunition storage magazines for indoor storage and berms and constructing pads for outdoor storage. Engineers may be tasked to supply appropriate dunnage for ammunition stacks according to DA Pamphlet 385-64.
- Constructing quarters and support facilities for ammunition facility personnel and security forces. This includes associated power and sanitation requirements.
- Constructing and maintaining perimeter security fences, lighting systems, or other required security, infrastructure protection and protection measures.
- Constructing fire breaks in and around the facility.
- Having staff proponency for the integration of environmental considerations and the linkage to the surgeon on matters of the included FHP issues.
- Planning for and installing adequate lighting protection systems.

PLANNING CONSIDERATIONS

12-13. Planners must consider a number of factors when they are designing ammunition storage and supply facilities, including drainage, shelter, ventilation, facility size, vehicle access, water supply, environmental impacts, adjacent property use, and facility protection:

- **Drainage.** Munitions can be damaged by excessive moisture and must be kept dry. Proper grading and, where possible, the installation of drainage facilities in the area of the ammunition facility will divert rainfall and groundwater away from ammunition stacks.

- **Shelter.** Ammunition and explosives must be sheltered from the elements and the enemy. Depending on the situation and the assets available, these shelters may range from approved steel, arched-earth mounded igloos, to an outdoor modular storage system reinforced with earthwork berms. These systems are discussed in detail FM 4-30.13.

- **Ventilation.** Adequate ventilation is required to protect stocks from moisture and to prevent the buildup of toxic and combustible gases.

- **Facility size.** The size of the facility depends on the kinds and quantities of munitions being handled. Facility size will be determined by the logistics unit commander, based on standards set forth in DA Pamphlet 385-64 and the tactical situation.

- **Vehicle access.** Vehicles that use the ammunition facility must be able to travel to and from the appropriate pickup points. Road networks and traffic flow patterns inside the facility must support concurrent resupply and issue operations and provide for the rapid evacuation of all vehicles in case of emergency. Firefighting equipment must have access to all parts of the facility.

- **Water supply.** Water tanks and reservoirs must be located to support firefighting activities. Refer to DA Pamphlet 385-64 for siting and resupply requirements and to FM 5-415 for firefighting procedures.

- **Environmental impacts.** With its associated FHP, environmental impacts are an aspect of protection that must always be evaluated and planned.

- **Adjacent property use.** Both present and potential future uses are important. Deeds of restriction may be required for adjacent properties in the vicinity of ammunition storage areas.

- **Facility protection.** The protection of an ammunition facility may be accomplished through a combination of facility hardening and dispersion and through active and passive security measures. These measures are similar to those described in the section on maintenance and supply facilities in this chapter. Generally, the ADC plan will stipulate what measures must be taken before, during, and after a damage incident and who will be responsible for each measure.

SITING AND LAYOUT

12-14. The logistics unit commander determines the locations of an ASP that best supports the scheme of maneuver and meets the commander's intent. The ASP is located within a reasonable support distance of the maneuver elements and, it is desirable to place the ASP near an established MSR (road or rail) to make stocking and distribution easier. However, ammunition storage facilities should not be placed too near major facilities, such as airfields, POL storage, and ports. Taking this precaution will reduce collateral destruction as a result of enemy targeting on other facilities. Level terrain with existing natural barriers and good drainage is preferable. This will serve to reduce earthwork requirements. If possible, existing facilities or structures suitable for conversion to storage areas should be used. The engineer advises the logistics unit commander on such matters as the location of construction materials, topography, drainage, and the condition of local road and bridge networks. Consideration must also be given to security and the ease of defense. Wherever possible, sites should provide a defilade to give concealment from enemy observation.

12-15. The specific layout of an ammunition supply or storage facility depends on the tactical situation, the terrain, and the type and amount of ammunition being handled. Engineers supporting the construction of ammunition supply and storage facilities advise the appropriate commander on construction and maintenance matters. If required by the tactical situation, the facility may have to receive and issue ammunition before construction operations are finished. Engineers may have to alter construction plans and techniques to allow for safe and efficient handling of ammunition while construction proceeds. Ammunition storage facilities are best arranged in dispersed storage areas. The separation of facilities provides protective dispersion; expedites the handling, receipt, and issue of materials; and facilitates inventory management and segregation. The road network is designed so that each area can be entered and exited independently. This prevents crossing traffic in all areas. Firebreaks wide enough (50 feet minimum) to prevent fires from spreading should be maintained. Soil that contains enough organic matter to allow it to burn must be excavated to the mineral subsoil. Since firebreaks around ammunition stacks are easily detected by aerial reconnaissance, their use may have to be restricted or camouflaged.

12-16.   Existing buildings may be used for ammunition storage as long as the rated floor load is sufficient. Chemical, incendiary, and white phosphorus rounds should not be stored on wooden floors since they are a fire hazard. Ammunition and explosives may be stored outdoors according to DA Pamphlet 385-64, which details site and layout requirements for the outdoor storage of ammunition. These supplies may also be stored on vehicles for adequate dispersion and rapid deployment.

12-17.   Special effects imposed by the local climate must be taken into consideration in the design and construction of ammunition storage facilities. The following should be considered:

- **Desert.** In the desert, the need for dispersion is extremely important since natural concealment is generally quite sparse. Shadows and regular-shaped patterns are conspicuous and can be avoided by the use of small, irregular stacks and the elimination of regular lines and rows. In this environment, engineers are seldom required to develop extensive road networks.
- **Cold-weather climate.** In a cold-weather climate, care must be taken to provide adequate dunnage for ammunition storage. Defilades must be avoided because they may be susceptible to flooding following a thaw. Engineer assets may be used to clear and maintain the road network in snow and icy conditions. See FM 5-430-00-1/Air Force Joint Pamphlet (AFJPAM), Volume 1.
- **Wet climate.** In wet climates, particularly the tropics, maximum effort must be made to combat the effects of moisture. Adequate shelter, dunnage, and ventilation must be provided.
- **Hot and cold areas.** In very hot and very cold areas, static electricity is an important consideration.
- **Drainage.** Drainage is an important consideration in certain climates and areas (including desert environments).

MEDICAL TREATMENT FACILITIES

12-18.   Regardless of the size, intensity, or duration of a conflict, medical treatment facilities (MTFs) are needed. With the fielding of Deployable Medical Systems (DEPMEDS) equipment to both the ISB and AO, construction requirements are greatly reduced. However, the requirements for site preparation remain high. The following are some of the GE tasks that must occur as part of the establishment of medical facilities:

- Site preparation.
- Trash and garbage pits.
- Soakage pits or a liquid disposal system.
- Incinerators (in addition to solid-waste requirements).
- Facilities such as showers, latrines, laundry services, food preparation, and dining.
- Water distribution.
- Hazardous-waste disposal.
- Regulated medical-waste disposal.
- Motor parking.
- An LZ.
- Perimeter security.

- Fuel storage.
- Power generation equipment placement.

12-19. The longer the anticipated duration of the conflict, the greater the need to support medical treatment through fixed facilities. While medical facilities always entail a considerable amount of environmental considerations in either temporary or fixed facilities, the importance of these considerations will tend to increase over time and should be considered and applied as early in the process as possible to minimize their effects over time. These facilities must have the capacity and degree of sophistication to treat injuries and other health problems sustained during the contingency. Design for a CBRN environment may be appropriate; it must promote rapid, high-quality treatment within the theater to expedite the Soldiers' return to their assigned duties. In addition, U.S. forces are responsible for the well-being of enemy prisoners of war (EPWs), DOD employees, contractors, and other nonmilitary personnel who accompany combat forces, such as the media and NGOs. The emergency treatment of multinational Soldiers or the civilian population may also be required.

HEALTH SERVICE SUPPORT SYSTEM

12-20. In the absence of DEPMEDS-equipped hospitals, consideration should be given to the use of existing facilities in the areas which were originally designed as MTF, or which are readily adaptable to use as an MTF. Attention should be paid to the types of buildings, their potential patient capacity, and their effective use before the conduct of an operation doing so may result in the selection or use of facilities that will save time and resources. The requirements for fixed facilities are generally restricted to the ISB, where hospital units do not move in conjunction with the redeployment of major tactical units.

12-21. The degree of permanence may range from a temporary field hospital, to a semipermanent station hospital, to the permanent construction of a general hospital. Site selection is the responsibility of health service support planners, who in turn, coordinate with the logistics staff officer. The logistics staff officer allocates the site and coordinates for the required GE support. These facilities should be located so that patients from the ISB can be easily brought in and safely transferred within the ISB from one medical facility to another. A location near ground transportation networks and close proximity to an air terminal is most desirable.

12-22. Hospitals may also be located to support high-density troop populations. MTF requirements are based on estimates of inpatient and outpatient loads and the theater patient evacuation policy. This policy establishes the number of days that patients may be held within the command for treatment. Then they either return to duty or convalescence or are evacuated to a facility outside the command. Shortcomings in major existing hospital facilities and all new requirements must be identified so that construction or rehabilitation can begin. Except when they are located in existing structures, general and station hospitals require many weeks for development before they can function. Once established, they can be moved only with substantial difficulty and time-consuming effort. TCMS or AFCS contains BOMs, estimates of man-hours of construction time, and plans for station and general hospital facilities and associated clinics.

SITE RESPONSIBILITIES AND PLANNING CONSIDERATIONS

12-23. The best sites have existing utilities, such as a potable water supply, sewage disposal, and electrical power. When new construction must be initiated, the site should be a relative geospatial high point and the subsoil should be free-draining. The site should be isolated and from areas where sanitation may be difficult and areas subjected to noise, smoke, odors, and other nuisances. It should, however, be located in an area that is conducive to expansion and safe for handling large volumes of fuels (up to 50,000 gallons of jet petroleum 4 (JP4) or diesel contained in collapsible fabric tanks). The fuel is needed for auxiliary power generation. The site should be located near waste collection facilities that can handle large volumes of contaminated material, including discarded food products and contaminated solids. The principles of phased construction will be enforced.

12-24. Lower-priority, complementary facilities may include a helicopter landing site, waste collection facilities, motor pools, laundry, vehicle parking, supply receiving and shipping facilities, and recreation areas. Even though waste collection facilities have low priority at the initial planning phase, the importance of them increases in direct proportion to the intensity and duration of the conflict since vast amounts of contaminated waste may be generated. Expedient methods for disposing of contaminated-waste products must be considered during the initial stages of planning. Such efforts must be designed to avoid any possibility of contaminating groundwater supplies. Expedient methods, whether they are landfill operations or incinerators, should be planned and located so that they enhance the operations of the medical facilities. These methods should also be planned for semifixed facilities, such as evacuation facilities like combat support hospitals (CSHs), to prevent them from contaminating their own groundwater supply, and potentially exposing patients and staff to infections. Provisions for a CBRN environment may be appropriate.

FACILITY PROTECTION

12-25. Precautionary measures taken to prevent or minimize damage as a result of natural disaster, accidents, and enemy activity are specified in the ADC plan. An MTF should not be located immediately adjacent to potential tactical targets, such as airfields, ammunition storage and supply facilities, POL storage, and major bridges. When the facility must lie within an established defensive position, it should be located away from the outer perimeter and at a distance from critical targets. The decision to camouflage a hospital or display the Red Cross© emblem rests with the tactical commander. All protection afforded to medical units under the articles of the 1949 Geneva Convention are compromised when MTFs are camouflaged.

INTERNMENT/RESETTLEMENT FACILITIES

12-26. Successful combat operations inevitably result in internees of some type. Internees fall into one of the following classifications:

- EPW, civilian internee (CI).
- Retained personnel (RP).
- Other detainee.
- Dislocated civilian (DC).
- U.S. military prisoner.

Note. Refer to FM 3-19.40.

Depending on the duration and extent of the conflict, the requirements for the evacuation of internees may warrant the establishment of internee holding areas within the corps area and semipermanent internment facilities within more secure locations. Further evacuation to semipermanent or permanent facilities outside the AO may also require provisions for total evacuation. The discussion of internment facilities in this chapter will be limited to the BCT, division, and theater levels. Generally, internees are evacuated for their own safety, for interrogation, for medical treatment, or to relieve troops in the capturing unit. Once internees are gathered at internment facilities, they constitute a pool of potential labor assets. They are, however, subject to special considerations and some limitations. Caution should be exercised when contemplating the use of internees of any type for labor, and the Judge Advocate General office should be consulted for guidance.

RESPONSIBILITIES AND PLANNING

12-27. The CCDR is responsible for I/R operations, and he provides engineer and logistics support to the military police commander for the establishment and maintenance of I/R facilities. Planning for the construction of I/R facilities must be developed early in the operational plan. This provides timely notification of engineers, the selection and development of facility sites, and the procurement of construction materials. The military police coordinate the location with engineers, logistics units, higher headquarters, and the HN. Failure to properly consider and correctly evaluate all factors may increase the logistics and personnel efforts required to support military operations. If an I/R facility is improperly located, the entire internee population may require relocation when resources are scarce. When selecting a site for a facility, the following should be considered:

- Locations where internee labor can most effectively be used.
- Potential threat from the internee population to logistics support in the proposed location.
- Threat and boldness of guerrilla activity in the area.
- Attitude of the local civilian population.
- Accessibility of the facility to support forces and transportation to the site for support elements.
- Proximity to probable target areas (for example, airfields and ammunition storage).
- Classification of internees to be housed at the site.
- Type of terrain surrounding the site and its conduciveness to escape.
- The distance from the MSR to the source of logistics support.

12-28. In addition, consider the—

- METT-TC.
- Availability of suitable existing facilities (avoids unnecessary construction).
- Presence of swamps, vectors, and other factors (including water drainage) that affect human health.
- Existence of an adequate, satisfactory source of potable water. The supply should meet the demands for consumption, food sanitation, and personal hygiene.
- Availability of electricity. Portable generators can be used as standby and emergency sources of electricity.
- Distance to work if internees are employed outside the facility.
- Availability of construction material.
- Soil drainage.
- FHP for EPWs and forces manning the site.
- Other environmental considerations as appropriate.

12-29. Engineer participation in managing internee activities includes providing construction support for building or renovating internment facilities and employing internee labor in engineer tasks where appropriate.

FACILITIES

12-30. The permanency and complexity of I/R facilities vary; however, internees must be provided areas that are dry, environmentally controlled, lighted, and protected from fire. They must also conform to sanitary rules, including the best practicable provisions for baths and showers. Internees must be allowed to exercise and have access to fresh air. Sexes must be segregated.

SITE SELECTION

12-31. Internment facilities must be planned soon enough in a contingency operation to provide for timely site selection and development. Construction materials must be procured and construction initiated promptly. Construction should be planned to maintain a standby capability for the acceptance of additional EPWs. The site should be located on a relatively topographic high point with free-draining subgrade soil. This will serve to minimize earthmoving requirements for drainage. Greater sanitary precautions must be taken when working with high-water tables or swamp-like environments. Planners should also ensure the availability of a potable water supply, a sewage system, an electrical power supply, and nearby supplies of construction materials. If possible, existing structures should be used to minimize new construction. The same basic safety and environmental considerations and conformity apply.

TYPES

12-32. Within locations in the AOR, internment facilities are classified as detainee collection points, detainee holding areas, or theater internment facilities. The facilities are built to contingency or enduring construction standards and vary in size, depending on the number and classification of internees. Under certain circumstances, semipermanent construction may be authorized. As with any TO construction, existing facilities that can be used directly or modified with a justifiable effort are preferable to new construction. Examples of internment facilities are shown in figures 12-1 through 12-5, pages 12-10 through 12-14. The degree of engineer effort required for the construction of these facilities varies from little or no support to a full-scale GE effort. Normally, the degree of engineer effort will depend on the permanence of the facility, the size and scope, and the availability of engineer assets.

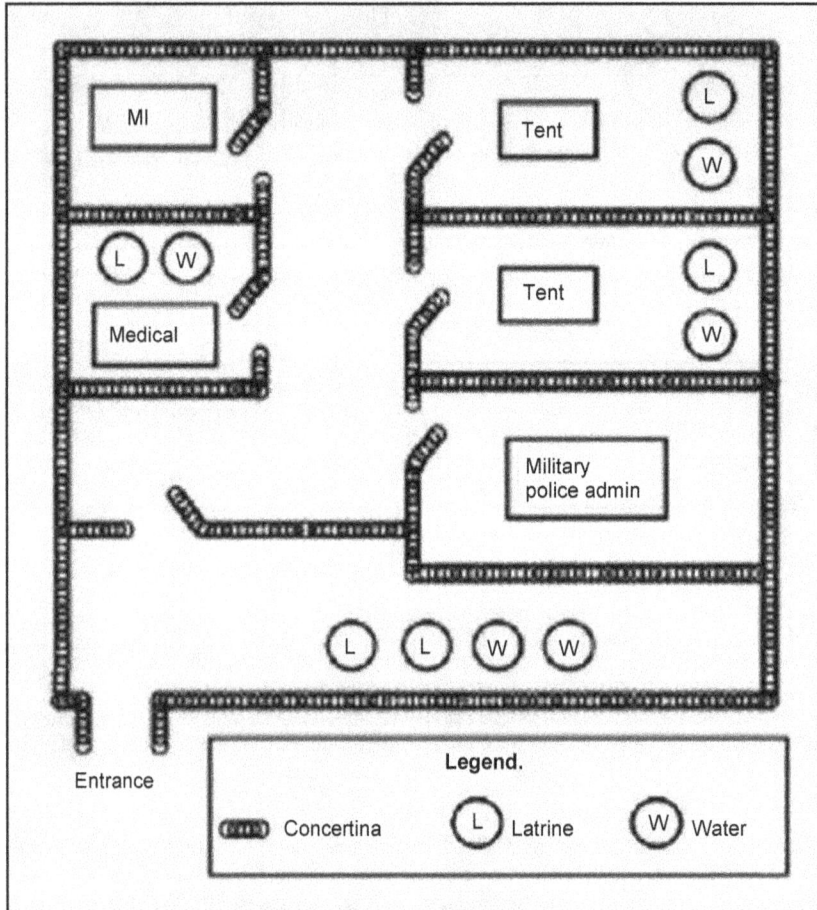

Figure 12-1. Sample detainee collection point

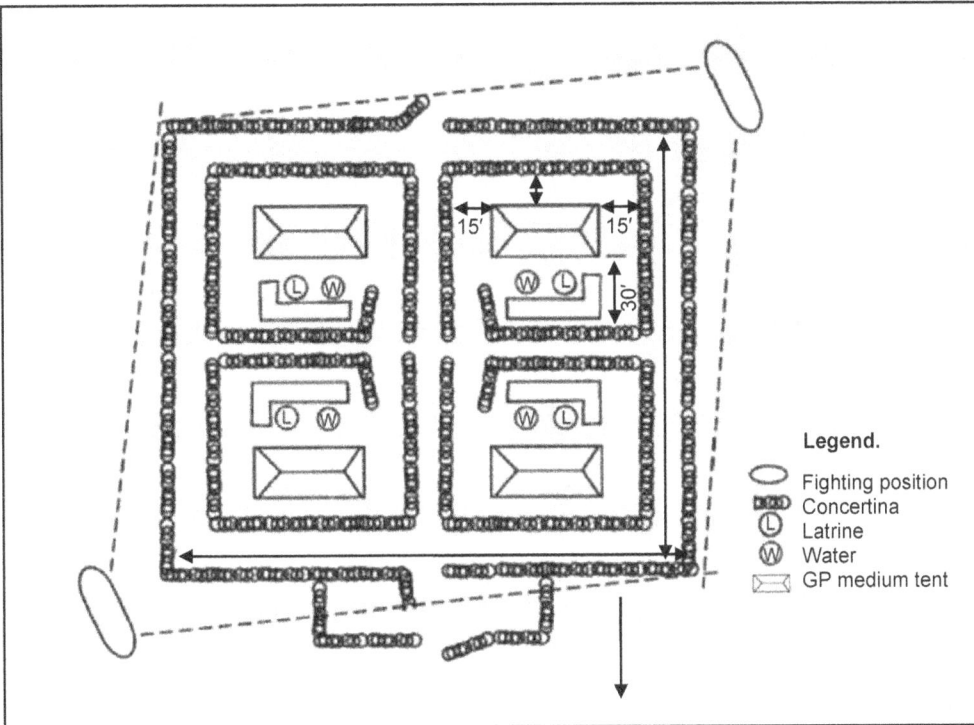

Figure 12-2. Sample detainee holding area

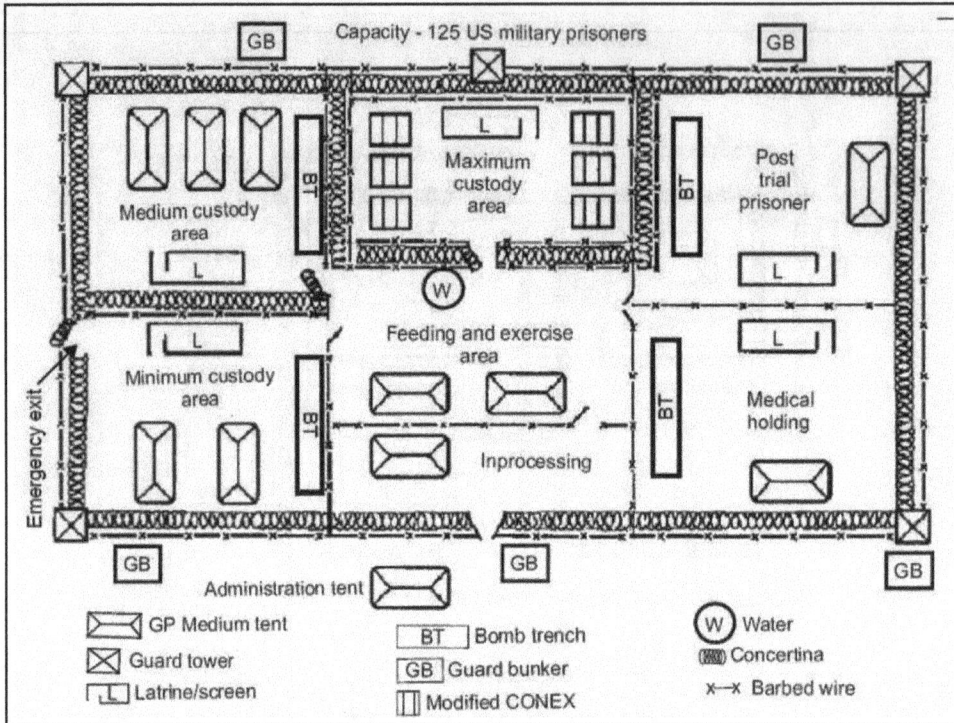

Figure 12-3. Sample field detention facility

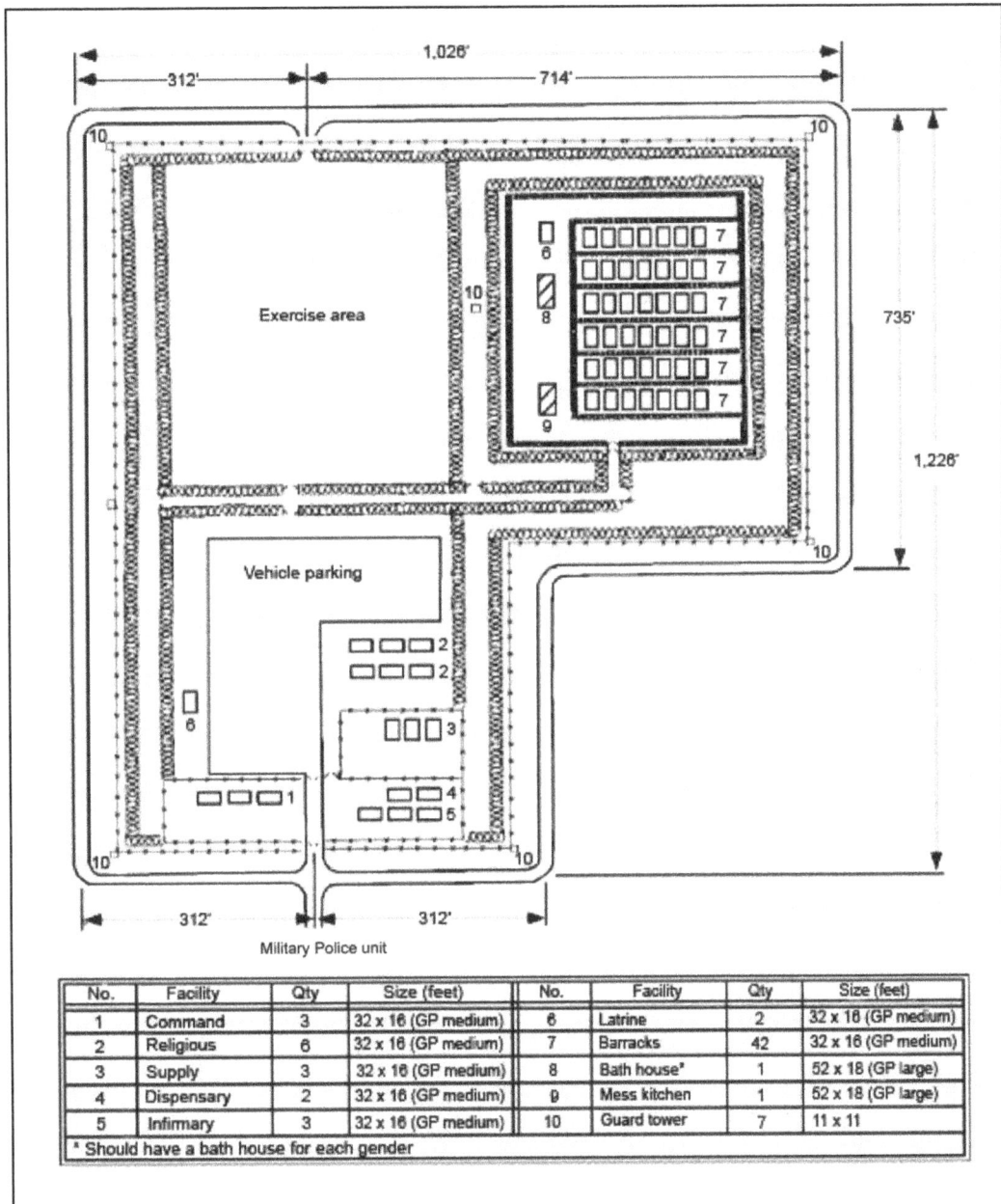

No.	Facility	Qty	Size (feet)	No.	Facility	Qty	Size (feet)
1	Command	3	32 x 16 (GP medium)	6	Latrine	2	32 x 16 (GP medium)
2	Religious	6	32 x 16 (GP medium)	7	Barracks	42	32 x 16 (GP medium)
3	Supply	3	32 x 16 (GP medium)	8	Bath house*	1	52 x 18 (GP large)
4	Dispensary	2	32 x 16 (GP medium)	9	Mess kitchen	1	52 x 18 (GP large)
5	Infirmary	3	32 x 16 (GP medium)	10	Guard tower	7	11 x 11
* Should have a bath house for each gender							

Figure 12-4. Sample 500-man enclosure

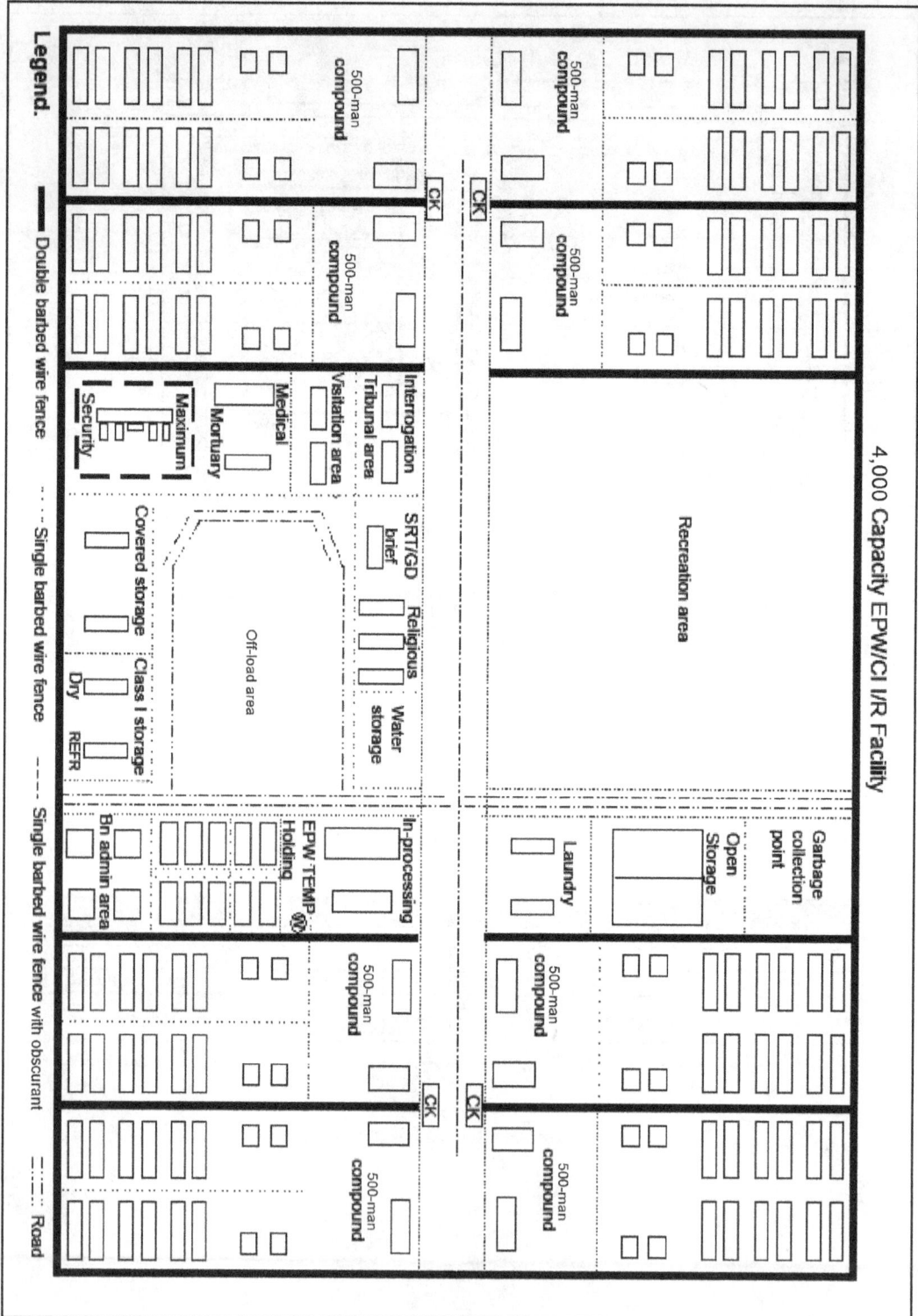

Figure 12-5. Sample theater internment facility

NEW CONSTRUCTION

12-33. Construction standards, BOMs, and estimates of man-hours of construction effort are contained in TCMS and the AFCS for I/R compounds. If facilities must be built, consideration must be given to the length of time that the facility will operate to assist in determining the standard of construction. Engineer support to the construction of I/R facilities may include the following:

- Security fencing and obstacles.
- Lighting.
- Towers.
- Vegetation-clear zone.
- Patrol roads adjacent to or outside the facility.
- Barracks, dispensary, mess, baths, and latrines, with related water and power facilities.
- EBS in conjunction with an EHSA for the site.

ENEMY PRISONER OF WAR LABOR

12-34. EPWs constitute a significant potential supply of both skilled and unskilled labor. How they are used is governed by those portions of the Geneva Convention relative to the treatment of prisoners of war (see FM 3-19.40). EPWs may possess engineer-related labor skills. The camp commander can assure the best employment for each EPW by establishing and maintaining occupational skill records. Approval for work on a project is obtained through operations channels. The use of EPW labor assumes a nonhostile attitude on the part of the EPWs. The commander, in deciding to use EPW labor, must weigh how essential the required work is against the personnel (security and support) and logistics effort required to provide the EPW labor. Generally, the significant effort required to manage EPW labor means that EPWs are only used in the absence of qualified local labor or contractors or when the commander determines that military engineers are not available or must be employed elsewhere. Prisoners of war should be used to the maximum extent for all work necessary in the administration, management, construction, and maintenance of EPW camps and facilities. EPWs—

- May not be retained or employed in an area subject to hostile fire in the combat zone. This generally precludes the use of EPWs forward of the communications zone (COMMZ).
- May volunteer, but may not be compelled, to transport or handle stores or to engage in public works and building operations which have a military character or purpose.
- May not be employed in labor considered to be injurious to health or dangerous because of the inherent nature of the work.
- May not be assigned to perform work considered as humiliating or degrading. This does not include those tasks required for the administration or maintenance of the EPW camp itself.

This page intentionally left blank.

Chapter 13

Real Estate and Real Property Maintenance Activities

Except when war is waged in the desert, noncombatants, also known as civilians or "the people," constitute the great majority of those affected.

Martin Van Creveld

Land and fixed facilities are needed within the AO to support committed forces. Included in this support is the acquisition of real estate for use as office space, billeting and housing, mess, material storage, staging areas, maintenance functions, training, ports, roads, buffer or safety zones, and so forth. These facilities may be used to house operations, planning, administrative, logistics, maintenance, and other functions. Existing facilities are used whenever possible for tempo enhancement. The engineer construction and other military support effort can then be invested in other immediate commitments. The Army's engineer real estate team and USACE CRESTs provide real estate support to U.S. forces both in CONUS and during contingency operations by obtaining land and facilities and managing the leases and use agreements. In the absence of existing facilities, new construction may be advised. The acquisition of privately owned property overseas, whether improved or unimproved, will almost always be accomplished through leasing. The use of HN land and facilities should be via written agreement. Whenever possible, integrate an EBS and EHSA into the process of obtaining real property and real estate. This is to ensure that EBS and EHSA meet acceptable levels of all appropriate inquiry into the previous ownership and that the uses of the property are consistent with good commercial or customary practice. The goal of this process is to identify environmental conditions that may present a material risk of harm to public health or the environment.

OBJECTIVES

13-1. The efficient conduct of real estate activities depends largely on a command-wide understanding of the objectives of the real estate program in overseas commands. These objectives are to—

- Acquire and administer real property essential to the mission.
- Acquire and use existing facilities to keep new construction to a minimum.
- Acquire environmentally safe real property and facilities that promote FHP, and coordinate for the performance of an EHSA when applicable.
- Protect the United States and its allies against unjust and unreasonable claims and charges for using, renting, or leasing real or personal property. Linking an EBS to the signing of a lease whenever possible is an excellent method of providing desired financial protection for the government.
- Provide reasonable compensation to individuals or private entities for the use of real property, except enemy-held property or, possibly, when such property is located in a combat zone or in enemy territory.

DEPARTMENT OF THE ARMY POLICIES

13-2. DA policy concerning real estate acquisitions is described in AR 405-10, TM 5-300, and DOD Directive 4165.6. Real estate operations in overseas theaters are based on the following general principles:

- **Adhere to international conventions.** U.S. forces will adhere to the provisions of the Hague Convention (1907), the Geneva Convention Relative to the Protection of Civilian Persons in Time of War (1949), the Hague Convention Relative to the Protection of Cultural Property in the Event of Armed Conflict (1954), and FM 27-10.
- **Conform to international agreements.** The Army real estate program will conform to international agreements and all other agreements affecting the United States, such as treaties, memoranda of understandings, land leases, and reciprocal aid, military assistance, SOFA, and CA agreements.
- **Make appropriate compensation.** When required, a fair and reasonable rental will be paid for real estate used, occupied, or held by the U.S. Army. Payment for the occupation of land will not be made to any person or persons who are members of the enemy government or who are hostile to interests of the United States. Compensation will not be made for any real property located in the combat zone, that is lost, damaged, or destroyed as a result of military action.
- **Honor HN laws.** U.S. forces will honor the real estate laws and customs of the host country to the fullest extent possible consistent with military requirements.
- **Use existing facilities.** U.S. forces will use existing facilities as much as possible to reduce the need for new construction and to conserve resources, time, and personnel.
- **Facilitate AT and other protection requirements.** U.S. forces will recognize, understand, and adhere to theater command AT and other protection requirements for facilities with regard to security and protection.
- **Minimize acquisition.** Real estate acquisition will be held to an absolute minimum, consistent with military requirements, to minimize disruption of the local economy. Joint utilization should be encouraged and the duplication of function and services should be avoided.
- **Follow appropriate acquisition policies.** The full use of the HN's governmental agencies will be made whenever possible, if not restricted by treaties. The acquisition of real estate in an overseas TO will be by requisition, by lease, or through consignment by the HN to the United States where the property is in the territory of an ally or by requisition, confiscation, or seizure when the property is owned by the enemy.

RESPONSIBILITY FOR REAL ESTATE

CHIEF OF ENGINEERS

13-3. The Chief of Engineers is the DA staff officer responsible for real estate functions and, as such, exercises staff supervision over Army real estate activities of overseas commands. The Chief of Engineers is responsible for carrying out the following duties:

- Provide technical advice and assistance in handling real estate acquisition, management of lease actions, and disposal.
- Issue instructions.
- Enforce applicable directives, policies, and regulations.
- Review records and reports.

COMBATANT COMMANDERS

13-4. CCDRs are responsible for all real estate activities within their AO. This responsibility may be delegated to a designated deputy or to the Army, Navy, or Air Force Service component commander with the greatest requirements and the appropriate authority and technical real estate staff expertise. Maintaining a single inter-Service real estate facility-use policy consolidates activities, reduces duplication, and limits the impact on the local economy. The theater commander may either establish a central real estate office to direct and record all real estate activities or direct that the commander assigned real estate responsibility establish such an office. Associated guidance on environmental considerations will be provided in annex L of the OPORD or OPLAN.

JOINT FORCE COMMANDERS

13-5. JFCs are responsible for carrying out the following duties:

- Determine real estate requirements.
- Plan, execute, and analyze real estate operations according to the pertinent directives, policies, and regulations.
- Prepare budget estimates and justifications as directed.
- Secure funding for lease payments from the appropriate using command.
- Prepare and submit real estate reports as directed.
- Conduct utilization inspection according to the instructions and criteria furnished by the Chief of Engineers.
- Notify the Chief of Engineers of utilization problems that require action at Headquarters, DA level.
- Furnish the Chief of Engineers with copies of all intercommand real estate and space utilization directives.

ARMY SERVICE COMPONENT COMMANDER

13-6. If the ASCC is assigned responsibility for real estate operations in the AO, all or part of this responsibility may be redelegated to the COMMZ commander. The acquisition of real property interests and lease management will be accomplished by qualified and delegated real estate teams or the CREST. The TA commander may retain control of real estate in the combat zone, redelegating responsibility for rear areas only.

ARMY SERVICE COMPONENT STAFF ENGINEER

13-7. The ASCC staff engineer operates and manages real estate and property acquisition, maintenance, and disposal functions. Duties include—

- Furnishing technical real estate guidance and advice to the commander, staff, and all echelons of command.
- Recommending real estate policies and operation procedures to the logistics officers.
- Preparing, coordinating, distributing, and exercising staff supervision over the execution of theater real estate directives with approval by the logistics officer.
- Acquiring, managing leases and use agreements, disposing of land or facilities, paying rents and damages, handling claims, and preparing records and reports for the real estate used within the AO.
- Maintaining an area real estate office in the AOR.
- Preparing long-range real estate plans and requirements.
- Using existing facilities as much as possible to reduce the need for new construction.
- Exercising staff supervision over real estate operations of subordinate commands.
- Ensuring compliance with international agreements and the law-of-land warfare.
- Coordinating with the authorities of the friendly HN.
- Ensuring the integration of appropriate environmental considerations in all real estate policies, operational policies, and operations.

13-8. The ASCC executes all real estate functions in the AOR when delegated such responsibility by the CCDR. When the commander of another Service component command is responsible for real estate activities, the ASCC engineer executes only ASCC real estate functions.

Subordinate Command Engineer

13-9. Engineers of commands below the ASCC staff engineer are responsible for furnishing technical real estate guidance to the commanders, staffs, and subordinate echelons of the commands. They handle other real estate duties as may be assigned or subdelegated to them by the TA commander.

Army Engineer Real Estate Team

13-10. The engineer real estate team or CREST, with the appropriate delegations from the Army Secretariat or the Chief of Engineers, is responsible to the area support command or other command as appropriate. They conduct real estate operations within their assigned areas according to the directives, instructions, and standing operating procedures (SOPs). Their duties include—

- Acquiring, managing leases and use agreements, disposing of land or facilities, and paying rents and damages for real estate used within the AO.
- Investigating, processing, and settling real estate claims.
- Conducting utilization inspections.
- Recording, documenting, and preparing reports on the real estate used, occupied, or held by the Army (or joint forces as appropriate) within their assigned areas.
- Coordinating with agencies of the friendly HN to execute joint U.S. and HN real estate functions.
- Coordinating with the SJA for legal issues and claim settlements.
- Including an EBS and EHSA, whenever possible and appropriate, in all real estate actions.

PLANNING

13-11. Real estate operations plans are based on directives or instructions issued to the CCDR by the Joint Chiefs of Staff (JCS) or by the Service commander appointed as executive agent for the JCS. The CCDR, based on directives and instructions issued by the JCS, establishes other policies.

13-12. A planning group that includes the combatant command staff and representatives of all Service commanders must initiate real estate planning in the preparatory phases of a campaign. The agency that will execute real estate operations when the campaign begins is organized at this time and should participate in all planning activities. In addition to plans for real estate operations during hostilities, consideration should be given to real estate requirements for the occupation period after hostilities cease. This may be most critical for those requirements that will be expected to be met by new construction such as base camps.

13-13. The site selection process is a joint effort consisting of several members (see FM 4-20.07 for more on-site selection team and site selection considerations). If available, a USACE CREST member should be an integral part of the site selection team, helping to ensure site acquisition through a HN use agreement or lease of private property. The site selection team should also include engineer, medical, or other required expertise to conduct an EBS and EHSA and integrate appropriate environmental considerations whenever possible.

13-14. Qualified personnel are essential for the handling of real estate responsibilities. Such activities can have major consequences in relations between U.S. forces and the HN. Military legal officers, USACE counsel, and civilian lawyers familiar with the laws of countries within the theater must be included to assist the planning group with advice and technical review of proposed real estate policies and procedures.

Property Acquisition

13-15. During conventional combat operations in the active combat zone, real estate required by U.S. forces is acquired by seizure or requisition, without formal documentation. Seizure is resorted to only when it is justified by urgent military necessity and only with the approval of the commander who has area responsibility. HN property may be occupied without documentation to the extent that tactical operations dictate and according to the U.S./HN agreements. After cessation of hostilities, private property will be leased if the property is still needed for 30 days or longer and if the property owner is known.

13-16. Normally, property is obtained through requisition, which is a demand upon the owner of the property or the owner's representative. No rent or other compensation is paid for requisitioned or seized property in the combat zone. This includes its use or its damage resulting from acts of war or caused by ordinary military wear and tear.

13-17. Outside the active combat zone, property is acquired only by requisition and lease and all transactions are documented thoroughly under the applicable provisions of theater directives. Large tracts of real estate are required for ports, staging areas, training and maneuver areas, leave centers, supply depots, base camps, and headquarters installations. Some of this property may be highly developed and have considerable value to the civilian population. The procedures must be followed in order to provide the required property, while ensuring that the legal rights of its owners are protected. Occupying units are responsible for providing funds for lease payments.

EXISTING FACILITIES

13-18. Existing facilities should be used whenever they are available. The advantages of using existing facilities are shown below:

- Swift occupation by military activities.
- Existence of utilities, telephone service, and connecting air, ground, and sea LOC facilities.
- Availability of on-site administrative and industrial equipment.
- Less diversion of troops from combat missions.
- Smaller outlay of government funds.
- Some inherent camouflaging of military activity.

13-19. The advantages of using existing facilities normally outweigh the disadvantages. Some disadvantages, however, may make facilities undesirable for military use. Planners should consider alternatives when existing facilities cannot be adapted to the desired survivability standards, when dispersion is difficult or impossible, when facilities cannot be tailored to meet military needs, or when environmental considerations and associated FHP issues make the site undesirable or questionable for occupation.

ACQUISITION

13-20. Local government officials can help identify available facilities or properties that meet or approximate military requirements. If these officials are unable to provide adequate assistance, military intelligence sources can be used to locate facilities. CA personnel and Army engineer real estate and/or CREST may work through local government officials or directly contact property owners to achieve agreements. Whenever possible, local government officials will normally evict and resettle any civilians from property requisitioned by the military forces. Only in the most urgent circumstances or upon refusal or inability of local authorities to act will U.S. military forces evict tenants or occupants.

13-21. A representative of the local government should assist in preparing all property inventories for local government-owned property. It is particularly important that requisitions carry the correct property descriptions and that local government officials check all requisitions against the corresponding entries in their permanent records. An EBS is also desirable to protect the government from future claims. If local records have been destroyed, the local authorities must establish a correct legal identification for the requisitioned property. The signature of the local official charged with real estate responsibility must be obtained on both the initial and release inventories. This official signature is required by international agreement to ensure that the U.S. government is protected from unjust claims for loss of or damage to property used by U.S. forces.

MODIFICATION

13-22. Instances may arise when it will be beneficial or perhaps essential to modify existing facilities to better serve military needs. Correcting deficiencies should be the primary focus of GE work. Theater planning should identify deficiencies and corrective actions that need to be taken. Theater requirements for AT and other protection tasks must be considered. AOR real estate principles for property acquisition apply as discussed above. Some additional compensation to property owners may be required, however. The ingenuity of Army engineers, HN and civilian contractors, combined with tools such as TCMS/AFCS, will be required to adapt existing facilities to military use.

FACILITY CONSTRUCTION

13-23. New construction in the AO is limited to facilities that are vital to the accomplishment of the overall mission, where no existing facilities meet the criteria required by commanders. The combination of land location and protection requirements may cause the construction of base camps to have a high likelihood of the need for new construction. See chapter 11 for a discussion of base camps.

REAL PROPERTY MAINTENANCE ACTIVITIES

13-24. RPMA are those actions taken to ensure that real property is acquired, developed, operated, maintained, and disposed of in a manner responsive to the mission. Acquisition, disposal, major and minor construction activities for new facilities, and additions or alterations to existing facilities are covered in chapter 12. This section includes operation, maintenance, and repair of facilities and utilities; fire prevention and protection; and refuse collection and disposal.

13-25. The RPMA function does not include the maintenance and repair of mobile and portable equipment or other items not classified as real property. Some of the coordination aspects of RPMA, however, do include many tasks not normally associated with minor construction and routine maintenance and repair aspects of RPMA.

RESPONSIBILITIES

13-26. RPMA is administered in the COMMZ by the theater Army area command (TAACOM) through its subordinate area support groups (ASGs). Support for RPMA is provided on an area basis to all installations, organic activities, and tenant units. The theater engineer command at the TA level provides overall supervision and technical assistance. The administration of RPMA forward of the corps rear boundary is a corps responsibility. Command relationships in the TA are described in FM 100-16.

PLANNING

13-27. The theater engineer command (or senior engineer brigade) and the responsible engineer staff must consider current and anticipated RPMA requirements for their AO. This will include—
- Maintaining and repairing the COMMZ.
- Estimating potential requirements for repairing war damage.
- Phase planning and target date requirements.
- Reviewing after-action reviews and lessons learned from recent operations.
- Considering contract support (such as logistics civilian augmentation program [Army]), preplacement of contract vehicles, and mechanisms for management of such contracts in theater (such as engineer advisory board or Defense Contract Management Command as in Bosnia).
- Considering other U.S. agencies (such as DOS and USAID) that may be in the AOR concurrently, and considering how the competencies of each might be leveraged. Establishing working relations across the agencies in peacetime.
- Determining the limitations (such as political) on using a cost-effective local work force and local contractors.
- Ensuring that a management system is in place that identifies facilities to support U.S. facility needs.

- Identifying procedures for accountability, security, maintenance, and training of appropriate local national facilities personnel if they are to be transferred to local authorities after the ceasing of hostilities.
- Integrating appropriate environmental considerations and related FHP concerns.

13-28. Any alteration or renovation work that is planned for existing structures should be designed according to the guidance of the AFCS and should essentially be of a nonpermanent nature. Plans for major repairs, renovations, or alterations on existing structures must include estimates for labor and materials. Planners may use estimating sources such as the engineer performance standards or a commercial estimating guide(means estimating guide).

13-29. There may be instances in the theater where the estimated materials or labor resources are in short supply or unavailable. Local materials and labor should be used to accomplish RPMA wherever possible. With the approval of the TA engineer, and with the support of the theater engineer command or the senior engineer brigade resources, the local engineer may change the design and/or scope of planned work to take advantage of locally available personnel and resources.

OPERATION OF UTILITIES

13-30. In the TO, the operation and maintenance or upgrade of existing utilities as well as the construction, operation, and maintenance of new utilities systems may be an engineer responsibility. Utilities systems include electrical generating and distribution systems, wastewater collection and treatment systems, and other special utilities systems, such as cooling and refrigeration, compressed air, and heating systems. Operating these systems requires specially trained personnel. They may be available through the theater engineer command (or the senior engineer brigade), trained locally, or hired from the local work force.

13-31. Since utilities systems must be reliable, measures should be taken to ensure their correct operation and to provide increased security if the situation warrants. Such measures include controlled access, continuous inspection, and adequate security personnel.

POWER GENERATION AND DISTRIBUTION SYSTEMS

13-32. If existing electrical generating and distribution systems are substandard or inadequate for military requirements, they will require upgrading or the installation of new systems. FM 3-34.480, FM 5-424, TM 5-683, and TM 5-684 give detailed guidance on installation, maintenance, and repair of electrical generation and distribution systems. Electrical supply in the AO can be accomplished in phases. Portable generating sets can supply minimum power requirements until fixed generation and distribution systems are installed.

Note. For more information on interior wiring see the National Electrical Code (NEC)® Handbook. For exterior wiring, see the National Electrical Safety Code (NESC)® Handbook.

WASTEWATER COLLECTION AND TREATMENT SYSTEMS

13-33. Large troop concentrations at fixed facilities generate requirements for sewage and wastewater collection and treatment. When existing fixed facilities are occupied, they usually include wastewater systems. However, these may not be operational or suitable for use by military forces. These systems should be operated, maintained, and repaired by engineer elements or qualified indigenous personnel. Construction, operation, maintenance, and repair of adequate sewage disposal systems are described in AR 420-1.

13-34. Field sanitation measures (such as pit latrines and grease sumps), portable chemical toilets, and waste treatment plants may be used temporarily until fixed facilities are completed and in operation. FHP is facilitated through good unit SOPs and field discipline in conjunction with the conduct of an EHSA and medical monitoring procedures.

13-35. As with all AFCS designs in the AO, the standard of construction for wastewater systems will nearly always be nonpermanent and designed to require minimum maintenance during the limited time anticipated for the period of occupation. Locally available materials may be used if approved by the ASCC engineer. Engineers will perform RPMA and operate the system as directed by the ASCC engineer. Guidance on environmental considerations will be provided in annex L of the OPORD.

OPERATION OF OTHER UTILITIES SYSTEM

13-36. In some areas, other types of central utilities systems may have to be operated by theater forces. These systems include heating, cooling, or refrigeration. Often, existing facilities will have utility equipment that must be repaired and/or maintained if it is to be operated. The responsibility for supervising this work will be directed by the ASCC engineer.

13-37. Local, portable, or unit systems (such as stoves and portable refrigeration units) will be maintained, repaired, and operated by the using unit. Engineers usually maintain central utility systems, such as steam plants, cold storage warehouses, or cooling plants. Where existing facilities are used, the theater engineer command may ensure the maintenance of these systems.

MAINTENANCE AND REPAIR OF FACILITIES

13-38. The maintenance and repair of facilities are the responsibility of the local commander, supported by engineer assets. Existing facilities that need maintenance and repair before they can be used are repaired to minimum standards. Repair materials must be estimated and prestocked to ensure that they will be available when needed.

13-39. Much short-term maintenance and repair work can be performed by local troops organized into self-help teams. These teams work with local logistics sources or supporting engineers to obtain the materials and tools they need. The early identification of spare parts requirements and the establishment of supply sources are critical. Adequately trained self-help teams can perform the majority of maintenance and repair work on their facilities, releasing engineer troops to accomplish more critical duties, complex repair work, and major construction projects.

13-40. When major repairs are required, the engineer unit assigned to the ASG, augmented when necessary with assets from the theater engineer command, makes repairs according to the priorities given by the TA engineer. Generally, the priorities are scheduled based on the impact the work has on the mission.

13-41. After immediate and ongoing maintenance and repair requirements are determined, a repair and maintenance program will be established using self-help and supporting engineer assets and/or local personnel or contracted support to accomplish the work. If the program is extensive or long-term, the unit commander should coordinate with the ASCC engineer to initiate a continuing facility engineer operation at the facility or installation. The facility engineer will then coordinate all requirements and resources needed to accomplish the mission. Further guidance on facilities maintenance and repair may be found in AR 420-1 and TM 5-610.

FIRE PREVENTION AND PROTECTION

13-42. Construction standards and materials in the AO make facilities very susceptible to fire damage and catastrophic loss of life or materials. TM 5-315 gives specific guidance for firefighting and rescue procedures in the TO. This TM prescribes the assignment of firefighting assets based on the supported population or facility area. For example, airfields, troop populations of 5,000 to 10,000 persons, or storage areas containing more than 100,000 square feet of storage space are each allocated at least one fire pumper truck team.

13-43. In all cases, and especially at smaller installations and facilities that do not have assigned fire protection equipment, the commander has responsibility for fire prevention and protection. All Army, command, and local fire regulations must be enforced. Programs of inspection must be established and self-help firefighting responsibilities assigned. Fire protection measures available to the commander include strictly enforcing the rules, setting up alarm and notification procedures, procuring and making available extinguishers and other firefighting equipment, and training personnel in fire prevention and protection measures. AR 420-1 provides further information about fire prevention and protection.

13-44. An additional requirement for the assets that provides fire protection is the requirement that they respond to HAZMAT spills. This support is an important part of environmental considerations that may have a direct effect on FHP.

REFUSE COLLECTION AND DISPOSAL

13-45. Improperly handled refuse can be a safety and health hazard. The local commander is usually made responsible for refuse collection and disposal. The command's engineers accomplish the task. Guidance on refuse collection and disposal may be found in AR 420-1 and NAVFAC MO-213/AFR 91-8/TM 5-634.

13-46. Landfill, burning, and removal are normal means for refuse disposal in an AO. Because of potential surface and groundwater contamination, the EH associated with uncontrolled methane gas production, increased vermin activity, and the obvious problems with refuse odors, it is imperative that landfills be properly designed and managed. Options available to lessen the quantity and/or eliminate specific types of refuse in sanitary landfills include incineration, recycling, composting, or a combination (see UFC 3-240-10A for more information pertaining to landfills).

13-47. Compaction and selective disposal are two other methods for reducing the volume of refuse. Selective disposal is the separation of certain types of refuse, such as wood or metal, from the refuse to be buried. The separated material is then stored or reused. Compaction is accomplished with specialized equipment for collecting and compacting refuse before it is dumped into the landfill. At the landfill site, special mobile compaction equipment may be used to reduce the volume of refuse before it is covered. Other compaction and refuse handling techniques include compacting and baling refuse for burial or removal from the area.

13-48. Refuse collection and disposal techniques depend on the volume of refuse to be generated, the duration of facility occupation, the presence of existing collection facilities, the resources available to perform the work, the area where the facility will be located, the situation, and the environmental aspects of the area. In some cases, selected recycling may even be enacted.

13-49. Special consideration should be given to hazardous waste, especially waste products generated by medical facilities and maintenance operations. Dispose of hazardous waste according to the appropriate regulations, laws, treaties, and agreements. Specific guidance should be contained in annex L of the OPORD. Improper disposal of such products may cause serious illness or death to those who operate landfills or cause irreversible damage to the environment. Specialized medical expertise exists to support the engineer and the commander when dealing with hazardous waste.

MILITARY REAL ESTATE OR REAL PROPERTY TRANSFER

13-50. The transfer of real estate or real property between units is an extensive process that can vary between commands and theaters. It requires planning and coordination between the current occupant and the incoming unit. If the transfer involves coalition or HN forces, there are special requirements to ensure that the United States is relieved of liability and that monetary reimbursements are made for any improvements (for more information, see the Base Camp Facilities Handbook).

13-51. During initial occupation, the inbound unit should focus on documenting facility conditions, inventorying the property, and reviewing all existing records. It is imperative that units properly manage and maintain records and reports as it will ease the transfer of future real estate or real property transactions with other units. Property agreements, the location, the description, the condition, surveys, the value, and maps are examples of items that should be requested from the outbound unit, kept on file, and updated as changes occur.

13-52. To complete a transfer, units must ensure that they have complied with all applicable laws and regulations before finalizing the transaction. Once the transfer is approved for acceptance or return, commands will—

- Coordinate with authorized representatives of units receiving the property, usually the real property accounting officer or designated representative (when feasible or required).
- Conduct a joint inspection of the property or facility, and prepare and sign a verification of joint inspection and record of return.
- Prepare and sign an outgoing inventory report.
- Prepare and sign a condition report.
- Prepare and sign an environmental summary report.
- Complete DD Form 1354 (Transfer and Acceptance of Military Real Property), ensuring that real property accountable officers or designated representatives of both organizations receive copies.

Chapter 14

Power Generation and Distribution

One should know one's enemies, their alliances, their resources and the nature of their country, in order to plan a campaign.

Frederick the Great

Electrical power is of critical importance as the military and the rest of society increasingly rely on electricity to conduct daily business and for basic life support. As the Army moves toward advanced digitization and automation to improve battle command, access to safe, reliable power has become an operational requirement. Likewise, utility-grade power is a component of the infrastructure that enables the execution and sustainment of military operations. In some cases, the level of electrical service available may serve as one of the primary measures of success for an operation. Consideration of electrical power requirements must occur from the lowest tactical level to the strategic level, across the spectrum of military operations and throughout the entire AO. Proper prior planning may prevent military and political issues from arising during the course of the operation. Since power is a component of GE, this consideration must include synchronizing work on the power system with the overall GE effort and associated environmental considerations during an operation.

RESPONSIBILITIES AND CAPABILITIES

14-1. Electrical power systems range from simple, unit-owned and -maintained tactical generators (TACGENS) to highly sophisticated multinational utility grids. Responsibility for each system varies widely, from large international corporations operating regional power grids, to individual units operating TACGENS for their command posts, to individual HN families operating small, portable generators inside homes. Each instance varies in complexity, efficiency, and reliability, but ultimately provides the user with a source of electrical power. As a theater matures, power requirements grow significantly, necessitating detailed updating of the plan by engineers to ensure that the power demand does not outstrip supply over time.

14-2. At the lowest unit level, organic TACGENS supply electrical power. Operation and maintenance of these simple power systems is a unit responsibility and does not require engineer support.

14-3. As the battlefield framework solidifies and the AO matures, the consolidation of clusters of small unit power systems is desirable. All but the most short-term, expeditionary facilities can benefit from centralization of the power system. The advantages of consolidated, centralized power generation and distribution systems are shown below:

- Reduced wear and tear on TACGENS, which are not designed for long-term, continuous operation.
- Streamlined maintenance and fueling demands.
- Increased cost efficiency per kilowatt-hour of power produced.
- Consolidated noise signature and security of power generation assets.
- More precise power with a higher degree of voltage and frequency control.
- Superior reliability of the power system.

14-4. Consolidation of power systems on a base camp, FOB, or other installation requires engineers to conduct detailed technical planning to best match projected requirements with available assets. Engineers conduct power planning in concert with overall facility master planning. For planning beyond the capability of organic engineers, specialized assistance may be obtained from engineer prime power assets (the 249th Engineer Battalion [Prime Power]) or by reachback to the USACE using tele-engineering or other means.

14-5. One, or a combination of the following, may achieve consolidation of power systems:
- Transition to commercial utility power.
- Centralized, contracted power plant with an interconnected electrical distribution system.
- Large deployable power generation and distribution systems (DPGDS) (up to 3.4 megawatts).
- Centralized military power plant (Army prime power or Air Force civil engineers).

Even after consolidation, transition to a more centralized power system never eliminates the need for the military user to provide redundancy by implementing a backup power plan using organic TACGENS.

14-6. At the operational level, electrical power is usually part of the national or regional infrastructure in the form of a commercial power grid. These power grids may be owned and operated by a government entity, a public or private corporation or a series of interdependent corporations. The sophistication of these power systems varies with the level of development of the location. In the most underdeveloped nations, the power systems may be small, crude, and well within the capabilities of engineer prime power units to operate, repair, and maintain. In more highly developed countries, systems may be highly complex with ultra-high voltage transmission, state-of-the-art dispatch and controls, and advanced generation methods. In these situations, engineer prime power units would be challenged to provide more than basic technical expertise because these conditions are outside their areas of core competence. Instead, rehabilitation of these systems will likely require the expertise of USACE electrical engineers, specialized contractors, large budgets, and long project timelines.

PLANNING

14-7. Electrical-power requirements and capabilities must be carefully integrated into the overall GE concept to support the objectives of the theater energy support plan. As with all niche engineer capabilities, the unit owning the capability is the SME and is best prepared to provide assistance in integrating and synchronizing their efforts with those of other engineer units. Requests for LNOs from prime power units and for staff assistance early in the planning process are crucial to achieving the desired effect with regard to power. When using existing power facilities, planning for the environmental risks associated with maintaining or upgrading a site needs to be considered. Recent experiences in Iraq have reminded planners and commanders of the effect of asbestos and other environmental problems on operations.

14-8. Planners should consider the full range of prime power capabilities when analyzing mission requirements and allocating units against those missions. Prime power units deploy, install, operate, maintain, and troubleshoot large power plants and distribution grids. However, their ability to provide technical expertise and battle damage repair of existing infrastructure produces more immediate and profound results than the time- and manpower-intensive process of constructing a new power system.

14-9. During complex missions (such as base camp construction), particular attention must be given to synchronizing prime power with the efforts applied to vertical and horizontal construction. Early involvement of the prime power unit in the master-planning process is crucial. It is important to have prime power personnel involved in regular construction meetings to maintain logical sequencing of the overall project, to minimize wasted effort, and to ensure that all parties involved in the development of the facility power system are working in a mutually supportive manner that supports the defined priorities for the project.

14-10. Engineer planners should set the conditions for mission success with the supported unit commander by managing expectations from the beginning of the planning process. Construction of a new power system is a laborious process with few interior tangible results until the system is complete and power is delivered. The same is true with repair or rehabilitation of an existing, battle-damaged system.

14-11. Planners should consider the development of the TO infrastructure in the terms of utilities, a skilled labor workforce, and sustainable power sources. Theaters with less developed infrastructures will require more prime power support than well-developed infrastructures. At wartime, however, developed infrastructures can be crippled in a short period of time. The extent of damage will influence the impact on the restoration of commercial power and may take months or years. Loss of commercial-power production will be detrimental to military operations and civilian activities. It will greatly increase the demand for electrical power produced by TACGENS, non-MTOE generators, and prime power plants.

14-12. Planners should consider distribution voltage and frequency. This is critical if plans call for using commercial power. In most cases where U.S. forces will maintain a long-term operational presence in a theater, the transition of military facilities to commercial power is a likely and desirable end state for power. Planners should strongly consider basing the theater electrical standard on HN voltages and frequencies. When voltage and frequency are not compatible with the intended use, power must be obtained from an alternate compatible source or, when possible, converted for compatibility. Planners must avoid dual-frequency circuits inside buildings. The close proximity of 50- and 60-hertz circuits can create a 10-hertz harmonic that can interfere with communications equipment and other frequency-sensitive electronic equipment.

ELECTRICAL POWER SYSTEMS

POWER GENERATION

14-13. Power is produced by generators that may be driven by diesel engines, gas turbines, boilers, or alternative sources. Generators are machines and, as such, are subject to mechanical or electrical failure. They require periodic maintenance and service to avoid breakdown. To obtain a source of continuous or prime power, multiple generators are installed in parallel. This arrangement allows the performance of maintenance on one or more generators while the others produce power. In developed countries, many large commercial power plants may be linked together to form an interconnected, redundant national or regional power grid. This power grid is the means of sending available power from the source to the user and is known as transmission or distribution.

PRIMARY POWER DISTRIBUTION AND TRANSFORMATION

14-14. Primary distribution networks carry medium-voltage power from the power plant to the transformers or substations. Primary distribution systems may be constructed with an extra-heavy-duty, multiconductor, shielded power cable that is suitable for ground-laid or buried applications, or they may be constructed overhead using poles and bare cable as part of an aerial distribution system. Transmission and primary distribution are in the medium, high, or ultra-high voltage ranges to efficiently and effectively send the power from the source to the user over long distances.

14-15. Army MOS 21P and 21Q prime power units are trained and qualified to work on distribution systems up to 33 kilovolts. Voltages beyond 33 kilovolts require specialized commercial personnel and protective equipment.

TRANSFORMATION

14-16. The medium-voltage power that is distributed on the primary system is stepped down to user level voltage by transformers. Most transformers are more than 95 percent efficient. As a result, very little energy is lost in the transformation process. The power put into a transformer approximately equals the power coming out. In the case of step-down distribution transformers, the high-voltage, low-current power going into a transformer approximately equals the low-voltage, high-current power coming out. When a transformer reduces voltage, it increases the current proportionally.

14-17. Primary distribution voltage (medium voltage) is stepped down to user level voltage by distribution transformers or substations. A primary distribution system may incorporate either or both of these items. Using distribution transformers allows power to be distributed at a higher voltage on smaller conductors and helps to reduce voltage drop and line loss. For military purposes, MOS 21P Soldiers are qualified to test and install transformers and test, install, and repair substations.

SECONDARY DISTRIBUTION

14-18.   Once the voltage is stepped down to the user level voltage at the transformer, the secondary distribution network carries the power from the transformer to the user. Secondary distribution systems are constructed with multiconductor cable when possible.

14-19.   In military applications, engineer prime power units (MOS 21P Soldiers) are responsible for the secondary distribution of a power system from the transformer to the main distribution load center (if one is used) or to the input of the main distribution panel box. Interior electricians (MOS 21R) in vertical construction platoons and utilities detachments construct the secondary distribution from the load center or main panel box to the user.

POWER SYSTEM CHARACTERISTICS

14-20.   The consideration of power system characteristics is important in determining the use of power for specific applications. Some power system characteristics can be altered to suit the needs of the user.

OUTPUT VOLTAGE

14-21.   Output voltage is the measure of the voltage at the output terminals of the power system. Large output voltage alterations can be made by using distribution transformers. Small output voltage changes can be made by adjusting the controls on TACGENS and prime power generators. Devices, such as voltage regulators, can be used to make small voltage adjustments to commercial power.

SINGLE- OR THREE-PHASE POWER

14-22.   Most alternating current (AC) power is generated as three-phase power. Single-phase power can easily be obtained from a three-phase source. Three-phase power is provided at three separate output terminals that share a common neutral terminal. The voltage difference between phases is the result of each being +120° out of phase with the other two. For example, on a user voltage level three-phase, four-wire system (fourth wire being neutral) with a measured voltage from any phase conductor to neutral of 120 volts AC, 208 volts AC is available by connection to any two-phase conductors. This allows for the connection of loads requiring 208 volts AC user voltage.

14-23.   Three-phase power systems should be designed so that each phase carries about the same amount of load as the other two. This concept is called "load balancing." Badly unbalanced loads will result in frequent tripping of protective devices and may damage equipment.

OUTPUT CAPACITY

14-24.   Output capacity is the amount of power a system can deliver. It is usually measured either in apparent power, kilo volt-amperes, or real power in kilowatts with an associated power factor. Output capacity is limited not only by the size of the generation equipment but also by the rated capacity of the distribution system. Electrical conductors and devices (such as transformers, breakers, and switches) are designed and manufactured with specific limitations on current and voltage. When the power demands of the user exceed the output capacity, the system is said to be overloaded and one of two things may occur. Either protective devices (such as fuses, breakers, or relays) are blown or tripped or else the system is damaged. The damage can occur in the form of melted conductors, burned connections, or blown transformers. Output capacity may be increased by upgrading distribution systems and by employing additional or larger generators.

RELIABILITY

14-25. Reliability is the measure of the ability of a power system to fulfill all the demands of the user without failure for long periods of time. Systems that are susceptible to outages, either scheduled or unscheduled, or that cannot provide all the power users need are not very reliable. Reliability can be improved by employing standby and load-sharing generators. It can also be improved by using redundant distribution systems (loop circuits) and enhanced by maintaining existing distribution systems and generation equipment.

PORTABILITY

14-26. Portability is the ability to rapidly relocate a power system that may be critical to certain operations. TACGENS are the most portable systems available. Since commercial power is tied to fixed facilities, it is the least portable. Prime power systems are portable, but require more effort and time to move and install than TACGENS. Prime power plant installation may be feasible if the plant remains in operation (stationary) for 30 days or longer.

14-27. AC power frequency is given in cycles per second or hertz. The most common worldwide systems are 50 and 60 hertz. The accepted U.S. standard is 60 hertz. Most countries establish one or the other as a national standard. They build their commercial-power systems accordingly. In a few countries, both systems may be encountered.

14-28. Some equipment is sensitive to AC frequency and will not operate properly when powered by a source with a different frequency. Units should ensure frequency compatibility for this equipment to avoid damaging it. Most transformers designed for 50-hertz operation can be used for 60-hertz application. Most 60-hertz transformers cannot be used for 50-hertz application unless they are significantly derated.

14-29. Prime power generation equipment can operate at 50 or 60 hertz. Most TACGENS operate at 60 hertz. Some specialized TACGENS operate at 400 hertz. The frequency power that is used extensively for aircraft systems, missile and avionics systems, signal systems, and some shipboard systems is 400 hertz. Frequency alterations are possible with the use of frequency converters.

LINE LOSS AND VOLTAGE DROP

14-30. Electrical conductors have some resistance. The amount of resistance depends on the type of metal, the cross-sectional area and length, and the temperature of the conductor. Copper is less resistive than aluminum. Conductors with larger cross sections and shorter lengths are less resistive than those with smaller cross sections and longer lengths. Conductors are less resistive at lower temperatures than at higher temperatures.

14-31. When electrical-current flows through a resistive material, some of the energy is converted to heat, causing a drop in voltage. The energy converted to heat is called line loss, and the drop in voltage is called "voltage drop." Distribution systems must be designed to safely carry the required amount of current while maintaining output voltage within the operating parameters of the devices being powered.

PRIME POWER OPERATIONS

14-32. Prime power operations are a subset of the GE function of the Engineer Regiment (see FM 3-34.480). Engineer prime power units provide an essential continuity between power from TACGENS and commercial sources (figure 14-1, page 14-6). Prime power units provide technical assistance and staff planning to support the development of electrical-power solutions for military operations. Prime power units also possess a limited organic capability to provide interim contingency power to satisfy the critical electrical requirements above the capability of TACGENS and below the availability of commercial power or to augment the power available from either source. The portion of the continuum that is exclusively prime power represents power generation and distribution accomplished by prime power units with their organic equipment. The intersections of TACGENS and commercial power with prime power represent areas of shared responsibility.

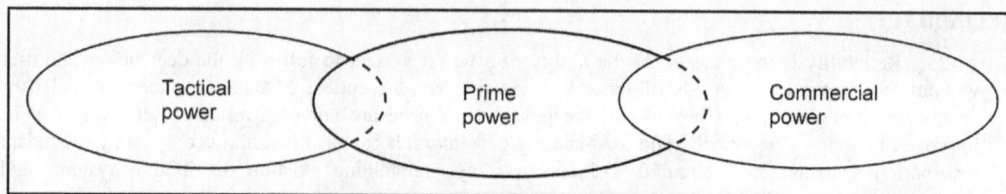

Figure 14-1. The power continuum

ORGANIZATION

14-33. The prime power platoon is the basic building block for conducting prime power operations. Prime power platoons are small, highly deployable, modular units that provide electrical-power support across the spectrum of military operations. The platoon is capable of deploying independently of its higher headquarters, but does require administrative and logistical support upon arrival in a TO.

14-34. If two or more platoons deploy for a mission, a prime power engineer company headquarters deploys to provide C2, sustainment, and specialized technical support to the mission. In peacetime, each prime power company consists of a company headquarters and four prime power platoons, augmented with one prime power platoon and one power line platoon from the Reserve Component.

14-35. These prime power companies are organized along with a headquarters and headquarters company under an engineer battalion (prime power). The battalion higher headquarters is USACE, which is an Army command. A cell from the battalion headquarters is deployed if more than one subordinate company is required to support a particular theater or contingency. This cell usually includes logistics, liaison, or coordination capability. It provides C2 of the companies; liaison and coordination; and specialized maintenance, administrative, and limited logistical support.

CAPABILITIES

14-36. Engineer prime power units support GE efforts theater-wide by providing advice and technical assistance on all aspects of electrical power. They provide limited, interim contingency power generation to critical facilities. This spans the spectrum of military operations, to include combat, stability operations, and homeland security. Prime power efforts and capabilities must be closely integrated and synchronized with the GEs effort to achieve the effects intended in a theater civil engineer support plan.

14-37. The prime power unit performs many technical, power-related tasks. A two-Soldier team will—

- Provide power-related planning and staff assistance.
- Conduct an electrical-load survey.
- Analyze and design power distribution systems.
- Perform damage assessment of distribution systems.
- Provide power-related technical assistance to the representative of the contracting officer.

14-38. Engineer prime power units can produce large quantities of reliable power with their organic generators. They can also install, operate, and maintain non-MTOE power generation equipment and some fixed commercial-power plants. This power generation capability can be used in a variety of military base camp configurations as well as seaports, airfields, C2 nodes, and other critical facilities. The power generation capability of the unit also allows it to operate, maintain, and perform damage assessments of fixed commercial diesel engine power plants. It uses organic or war reserve equipment to provide power to locations where another source is not available or is inadequate.

14-39. Each prime power platoon is equipped with four 1,050 kilovolt-ampere power units, giving the platoon 2.52 megawatts of continuous power production capability and 3.36 megawatts of peak power production.

ORGANIZATION EMPLOYMENT CONSIDERATIONS FOR PRIME POWER ORGANIZATIONS

14-40. The following guidelines will enhance the employment of prime power assets and will result in more reliable electrical service:

- Determine how much power is needed and the power source.
- Plan to upgrade service after initial installation.
- Determine the required level of service and reliability.
- Specify the date and duration of the requirement.
- Coordinate funding requirements.

14-41. The prime power unit will conduct a preliminary reconnaissance before committing assets. The prime power platoon can identify the power needs and recommend the best way to fulfill them. The platoon will conduct a load survey to determine how much power is required and where it is required and then design systems to provide power based on the survey. The prime power unit will recommend the best power source based on the level of reliability required and available assets. Many times, the power requirements are so complex that the supported unit is unable to communicate its power needs. A thorough reconnaissance will clarify their needs.

14-42. Commercial power is used when it is available. Commercial power is usually reliable in developed countries. Prime power platoons can make connections to commercial distribution networks or coordinate with the utility company to have them make the connection. Once connected, the system can provide continuous power service virtually maintenance-free. A major advantage of using commercial power over installing a plant is that the prime power platoon remains available to perform other electrical work. When a plant is installed, the platoon or part of the platoon is fully committed to operating and maintaining the plant instead of performing other power-related missions. This takes greater advantage of the technical training of the platoon.

14-43. The power source should be matched to the load requirements. Resources that are ill-suited for a particular application should not be committed. A common violation of this guideline occurs when a large prime power plant is installed to provide power to a relatively light load. This is an inefficient use of power generation assets that could be better used elsewhere. Operating large prime power generators under light loads also increases the wear and tear on generator engines. Prolonged misuse will cause carbon fouling and buildup, reduced engine performance, and eventual engine failure. Prime power equipment should be considered when the assessed load exceeds 437.5 kilovolt-amperes. However, prime power assets can be used for smaller loads in circumstances when reliable, continuous power is critical to mission accomplishment.

14-44. Load increases should be considered during planning and made to provide adequate power. If future plans indicate that growth will increase power demands, build distribution systems to handle the growth. This can be done either by overbuilding the system initially or by building it so that it can be readily expanded as needed. Systems that are not anticipating growth should still be designed and built to accommodate 150 percent of the estimated demand.

14-45. Plant deployment, installation, and distribution system construction is a time-consuming procedure. This process precludes the rapid relocation and setup of power plants and their associated distribution networks. Generally, it takes a full prime power platoon up to 5 days to construct one organic power plant and have it operational, depending on the amount of site preparation required. The 5-day rule of thumb does not include construction of a distribution system to provide power from the plant to the user.

14-46. Distribution system installation can take days, weeks, or months depending primarily on the size of the system to be constructed. Distribution system construction and installation is influenced by the following variables:

- The type of system required (underground or aerial).
- The availability of the BOM.
- The availability of other engineer assets for trenching or mine clearing.
- The threat of the enemy and local security situation.

- The availability of interior electricians (MOS 21R) to construct the secondary (interior and low-voltage) portion of the distribution system.
- The environmental considerations that may constrain construction or renovation of existing systems.

14-47. Generally, it is feasible to install a prime power plant for units or activities that plan to use it for 30 days or more. Units relocating often should use TACGENS or relocate to facilities powered by a commercial grid or an existing prime power plant. Expanding an existing prime power plant and its distribution network is usually more practical than relocating it.

14-48. Deployed units will rely on their TACGENS for initial power needs. Units and activities that are in place for extended periods will need to upgrade their facilities. Power produced by low-voltage TACGENS should be replaced by prime power or commercial power when units anticipate remaining in place for more than 30 days and load demand justifies the replacement. This replacement increases reliability and saves wear and tear on TACGENS. Stand-alone prime power plants should be replaced with commercial power as it becomes available. Prime power plants may be used for up to 6 months as a temporary power solution. At this time, supported units should plan to switch to commercial power, purchased commercial generators, or contract power assets (such as the logistics civil augmentation program, local contractors, and USACE contracts). When considering purchased commercial generators, the maintenance and refueling requirements must be addressed. Because of the unique capabilities and limited equipment of prime power units, organic prime power equipment should be used as a temporary power solution until a more permanent power capability is attained when supporting missions of a long duration. As a strategic asset, prime power equipment and personnel must establish power quickly and then support the next high-priority mission as the situation continues to develop. The desired end result is usually to use power from the highest level of the power continuum.

14-49. The priorities of employment of prime power support are the same as those for other engineer support in the TO. FM 5-116 and FM 100-16 list engineer support priorities in the TO.

14-50. Planners should consider the use of prime power war reserve assets when it is impractical to employ organic prime power equipment assets or when the mission is known to be long-term. Prime power Soldiers will install these assets and train the supported organic personnel (normally MOS 52D, power generation equipment repairer) to assist in the operation and maintenance of the plants. This provides the supported unit sustained reliable power, while reducing the long-term manpower requirements on the prime power battalion. War reserve assets may be used for backup power as well.

MULTISERVICE CAPABILITIES

14-51. The Navy has the construction battalion maintenance unit (CBMU) that provides follow-on public works operations to maintain and repair existing advanced base shore facilities or facilities constructed by NMCB during contingency operations. The unit is capable of equipping, manning, and maintaining steam and electrical power generation and distribution systems for advanced base facilities of up to 5,000 personnel. For more information on CBMU and NMCB capabilities, see Naval Warfare Publication (NWP) 4-04.1/MCWP 4-11.5.

14-52. The Navy also has mobile utility support equipment (MUSE), which provides power plants, substations, steam plants, and technical expertise to support DOD utility shortfalls worldwide. MUSE technicians, like their Army 21P counterparts, attend the prime power production specialist course and an additional MUSE familiarization course. Technicians are deployable within a 24-hour notice to provide technical assistance for organic and inorganic utilities. Technicians are capable and equipped to install, repair, maintain, and operate power generation, electrical distribution, transformation, and steam-generating equipment and infrastructure. MUSE teams are compact and multidimensional with utility skills that are applicable in nearly any utility situation. For more information on MUSE, see Naval Facility Command Instruction (NAVFACINST) 11310.2E and Chief of Naval Operations Instruction (OPNAVINST) 11300.5B.

14-53. The Air Force has both Prime BEEF and RED HORSE units that are deployed in UTC sets. Prime BEEF are mobile assets deployed to air bases, where they can combine to form a MOB combat engineer force of 200 to 320 people, depending on the threat and number and type of aircraft. RED HORSE units provide heavy repair capability and construction support when requirements exceed normal base civil engineer capabilities and where Army engineer support is not readily available. They are stand-alone squadrons that are highly mobile, largely self-sufficient, and rapidly deployable. Both of these units are capable of providing electrical-system installation.

14-54. Marine engineers are organized to accomplish specific tasks of limited duration focusing on support to Marine air-ground task forces (MAGTFs). They are capable of providing mobile electric power through their engineer support battalion (ESB), primarily to the MAGTF.

This page intentionally left blank.

Chapter 15

Petroleum Pipeline and Storage Facilities

...before the shooting begins. The bravest men can do nothing without guns, the guns nothing without ammunition; and neither guns nor ammunition are of much use in mobile warfare unless there are vehicles with sufficient petrol to haul them around.

Field Marshal Erwin Rommel, World War II

The noncontiguous battlefield of today is even more dependent on petroleum products. In the European Theater during World War II, about half the total logistical tonnage was petroleum fuel. During the Korean and Vietnam wars, this figure rose to about 60 percent. The battlefield of the future anticipates even greater consumption of these products. In the conceptual plan for supplying needed fuels, bulk petroleum is delivered through ports or LOTS. There, it is off-loaded into storage facilities and shipped forward. The modes of shipment in descending order of priority are pipeline, inland waterways, rail, motor carriers, and aircraft. The preferred method of shipment is by pipeline because pipelines save time, money, and resources for other logistical operations. Each of these methods has its own security considerations and risks that must be addressed in the planning process. The engineer mission is to provide general and specialized assistance in constructing and maintaining pipeline systems. The environmental considerations surrounding petroleum and petroleum products are huge; and even in the midst of operations, spills can have an effect that impacts operational commanders due to the volume of petroleum involved in spills, being set afire, or subjected to some other negative situation. FHP, environmental protection, and other related risks must be addressed in the planning for and operational activities surrounding bulk petroleum.

RESPONSIBILITIES

15-1. The joint petroleum office (JPO) under the logistics directorate of a joint staff (J-4) coordinates the petroleum needs of all services within the CCDR's AOR. The petroleum group commander for the ASCC is responsible for all aspects of petroleum distribution planning and related supply operations. The group performs liaison with the theater support command materiel management center and HN staffs for coordinating multinational petroleum distribution support. It distributes fuels based on priorities established by the ASCC and by directives received from the theater support command materiel management center.

15-2. The petroleum pipeline and terminal operating battalions distribute bulk petroleum in the AO. These battalions are responsible for the operation and organizational maintenance of petroleum pipelines and storage facilities. They are responsible for installing collapsible tanks and associated equipment for the tactical petroleum terminal (TPT). They also install collapsible hose lines used to temporarily extend pipelines.

15-3. The theater engineer command is typically the senior engineer headquarters that supports the petroleum distribution effort. See figure 15-1, page 15-2. The theater engineer command provides maintenance (excluding organizational maintenance) and repair of existing pipelines. It also designs, constructs, and expands the tactical pipeline system (including marine terminals and storage facilities).

These tasks are typically done by U.S. engineer forces (could also be multinational engineers or contracted capabilities) or through coordination with the HN.

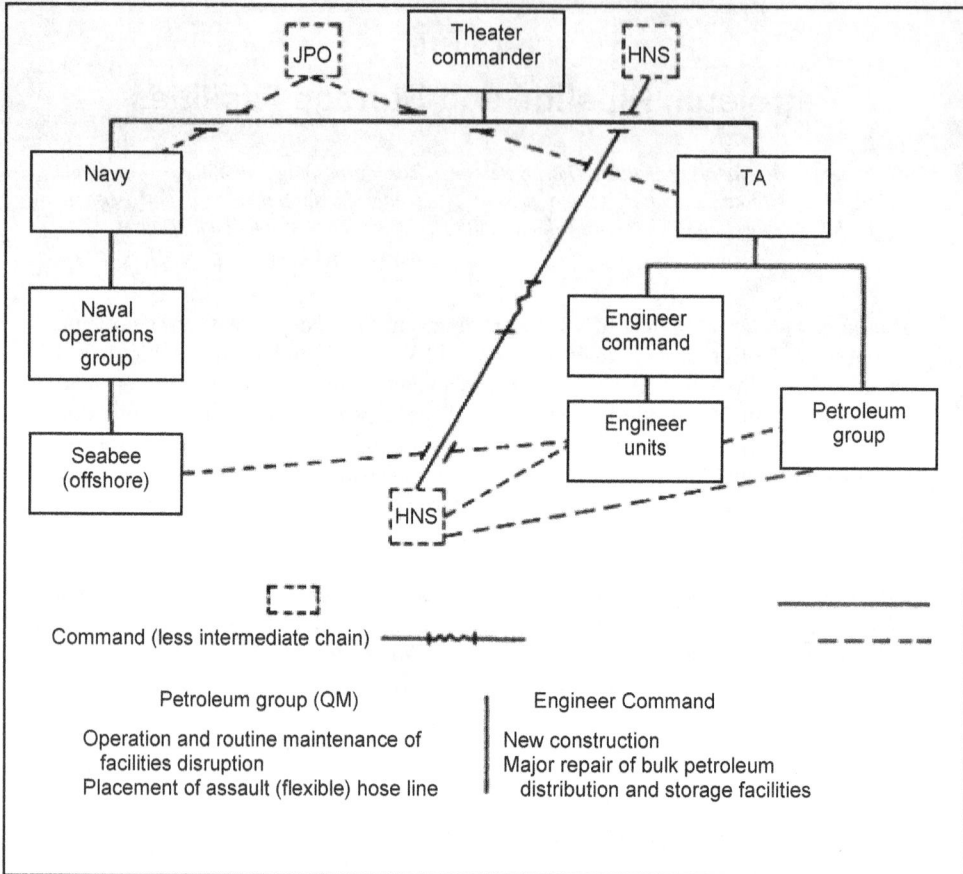

Figure 15-1. Engineer support to POL facilities

CAPABILITIES

15-4. Engineer support to the petroleum distribution effort calls for a combination of general and special construction skills. To maximize potential and minimize the duplication of low-density skills and equipment, GE construction units are augmented with specialized pipeline engineer units from the Active or Reserve Components. Each of these units must be well trained, possess experts knowledgeable of the environmental considerations associated with HAZMAT, and know how to minimize/mitigate the associated risks. They must also have the knowledge of how to integrate AT and other protection considerations into the design and operation of pipeline operations to minimize those risks.

ENGINEER HORIZONTAL CONSTRUCTION COMPANY

15-5. The primary military engineer unit required to support the pipeline construction company with the petroleum distribution effort is the horizontal construction company, augmented with a construction and geodetic survey design and material analysis section. These companies provide horizontal construction support for most of the tactical pipeline construction missions. Tasks include route repair and construction, gap crossings, pipe supports, storage tank erection, and pump station and dispensing facility construction. A unit with heavy earthmoving equipment can best do these tasks. Contractors may perform these tasks as well. The horizontal and pipeline companies may also provide supervisory personnel for HN employees that assemble pipe and associated equipment.

ENGINEER PIPELINE CONSTRUCTION COMPANY

15-6. These units provide technical personnel and specialized equipment to support up to three horizontal construction companies. They support construction by assisting horizontal companies in the construction, rehabilitation, and maintenance of pipeline systems. These units have a limited capability to construct, rehabilitate, and maintain pipeline systems without the assistance of horizontal companies.

MILITARY BULK PETROLEUM DISTRIBUTION SYSTEMS

15-7. The Army has used large-scale petroleum pipeline distribution systems since World War II. During World War II and shortly afterward, the total military pipeline system became standardized. Standardization included the bulk fuel distribution equipment. This equipment remained largely unchanged until the mid-1980s, when a major upgrade of materials and equipment took place. The entire distribution system is now subdivided into offshore and inland systems. The basic characteristics of each system and some of their prominent features are shown in figure 15-2 and are described in the following paragraphs. More in-depth characteristics and construction standards can be found in FM 10-67 and FM 10-67-1.

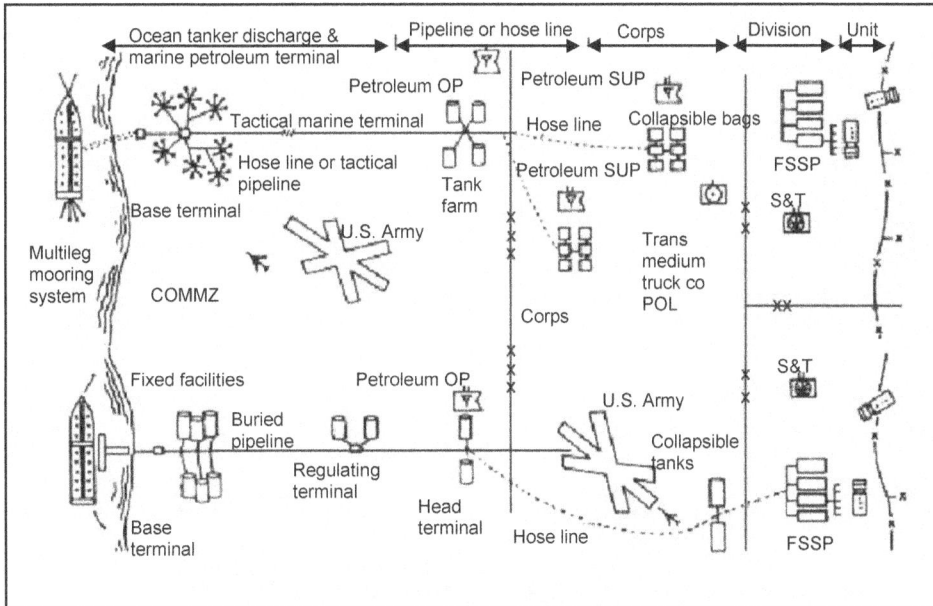

Figure 15-2. Example bulk petroleum distribution system

NAVY UNDERWATER CONSTRUCTION TEAM

15-8. An underwater construction team (UCT) is a specially trained and equipped unit that provides underwater engineering, construction, repair, and inspection capabilities to meet Navy, Marine Corps, or joint force operational requirements. In addition to many other capabilities, a UCT can repair underwater pipelines.

OFFSHORE PETROLEUM DISTRIBUTION SYSTEMS

15-9. The Offshore Petroleum Distribution System (Navy) (OPDS) is a set of equipment and material used to move petroleum from ships or barges to the first storage facilities on land. The OPDS may be installed entirely by U.S. Army engineer units or in conjunction with Navy construction units, depending on the specific situation. Both Army engineers and Navy construction forces have the capability to extend underwater pipeline up to 4 miles from the high-water mark. Such lines are needed where shallow waters or blocked channels prevent tankers from entering ports. If tankers can enter and use existing port facilities, engineers install fuel-unloading equipment at the pier or wharf. The first major storage facility is usually located within a 5-mile radius of the beach. Primarily, Army engineers and QM units are responsible for the construction of the beach terminal unit and the associated road networks. The Navy construction forces ensure that the OPDS is constructed properly and ties into the Army's beach terminal unit. JP 4-03 addresses this relationship in more detail. For guidance on the environmental considerations associated with bulk fuel, spills, and other offshore-related issues refer to NWP 4-11 and OPNAVINST 5090.1B.

TACTICAL PETROLEUM TERMINALS

15-10. TPTs have recently been developed to take advantage of new, rapidly emplaceable, flexible storage tanks. The standard TPT uses 18 of these 5,000-barrel (210,000-gallon) collapsible tanks to provide fuel storage. When the TPT is deployed at its maximum size, it requires an area of about 160 acres. The tanks are interconnected, filled, and emptied by a system of flexible hoses and trailer-mounted pumps. The petroleum pipeline terminal operations (PPTO) battalion is responsible for emplacing fuel tanks, hose lines, and pumps. This is not typically an engineer responsibility, although selected engineer support may be required.

15-11. Substantial engineer effort may be needed to help the petroleum operating battalion prepare the TPT site. The areas around the tank sites must be cleared of vegetation, and the sites must be leveled. Earth berms must be built to provide added support and horizontal protection for the tanks. The tank farm area must be properly drained to prevent water damage and to minimize problems from fuel spills or catastrophic tank failure. Interconnecting roads are needed within the tank farm as well as access roads and parking areas for heavy vehicles at fuel-dispensing points. A water supply for firefighting may need to be developed.

BOLTED STEEL STORAGE TANKS

15-12. Bolted steel tanks with storage capacities of up to 10,000 barrels (420,000 gallons) are still in the supply system. These tanks are especially useful at petroleum terminals in places where area restrictions preclude the optimum spacing of collapsible tanks or where more permanent facilities are required. The erection of the bolted steel tanks requires considerably more time and engineer effort than collapsible tanks.

INLAND PETROLEUM DISTRIBUTION SYSTEMS

15-13. The inland petroleum distribution system (Army) (IPDS) is a system of pipelines, hose lines, and storage containers that extends from the shore or port as far forward toward the combat area as required. The system consists of one or more main or trunk pipelines and pumping stations that move the product through the line, intermediate tank farms, and branch lines to large users (such as airfields) and the head terminal at the end of the line. The main pipeline may be an existing civilian pipeline provided by the HN, a line captured from the enemy, a tactical military pipeline constructed by military engineers, or a combination of these. The construction materials used in tactical military pipelines are easily assembled and readily adaptable to existing conditions.

MILITARY INLAND PETROLEUM DISTRIBUTION SYSTEMS PIPE AND COUPLINGS

15-14. The new standard pipe used in the IPDS is a 6- or 8-inch nominal diameter, coupled aluminum pipe. The pipe comes in standard lengths of 19 feet. The pipe ends have special grooves rolled on the ends to allow sections of pipe to be joined with a gasket and coupling. The couplings for the pipe are designed to close with a lever. The pipe is considerably lighter than the older steel pipe and tubing and can be joined much faster and with fewer people. Aluminum pipe comes with curved elbow sections, which allows the pipe to negotiate turns better. Aluminum pipe can be cut and the ends prepared in the field with special tools found in the IPDS inventory.

PUMPS AND PUMP STATIONS

15-15. Pump stations are located along the pipeline to maintain the pressure required to move liquid fuel. PPTO battalions operate the pump stations. These crews operate pumps, maintain equipment, and perform pipeline patrol between adjacent pump stations. Security patrolling may require additional personnel to accomplish. The spacing of the pump stations will depend upon the hydraulic design of the pipeline, as determined by the senior engineer headquarters controlling the engineer effort for the petroleum operations and by the anticipated future requirements of the system. On relatively flat terrain, pump stations will be about 15 to 20 miles apart. In mountainous terrain, pump stations may be much closer.

15-16. Pump stations consist of a set of pumps, station fuel storage tanks, various pipeline operating equipment, and personnel facilities for the crews. The tactical and logistical situation will dictate the other features of the station, to include any specific protection and AT requirements. The pump station should be located on relatively high ground to allow fuel vapors to move away from the facility. Personnel facilities should be located away from the operating equipment because of noise and the presence of noxious fumes. These and other considerations must be integrated into the engineering support plan. Involve medical personnel to do an EHSA to verify the level of FHP risks associated with these operations. Ideally, medical personnel will be included in planning to minimize the likelihood of these risks.

15-17. Assembling pump station components requires the specialized skills of personnel from the pipeline construction company. Newly introduced equipment significantly reduces construction time, because many of the components are modularized. Some fabrication, however, is still required.

PIPELINE CONSTRUCTION AND MAINTENANCE

PLANNING PHASE

15-18. The engineer planning phase for the construction of a petroleum pipeline begins as soon as the need for a pipeline has been established. The theater engineer command, in conjunction with the other subordinate controlling headquarters, determines the general route for the pipeline, incorporating appropriate environmental considerations. This ensures that the material required can be available when needed and that the environmental considerations of environmental protection and FHP have been integrated. In some cases, pipe has to be manufactured and shipped to the designated area. This may add months to the construction schedule. Proper planning can reduce some of these potential delays.

15-19. Early determination of required construction units and support must be made. Transportation needs must be well planned because engineer companies have a limited lift capability to move themselves or the required materials. The requirement to transport large volumes of pipeline material may also prevent rapid installation of the pipeline, but again, proper planning can reduce much of this potential delay.

15-20. Final selection of the pipeline route begins after a physical reconnaissance of the areas to be crossed is completed. The pipeline route will have the following major characteristics:

- Route follows secondary roads to reduce disruption of traffic on the MSRs.
- Route should be the most level ground available and avoid sharp changes in elevation. Pipeline supports and suspension bridges allow the construction of the line over small and large gaps, but add to the construction time and amount of additional construction material required.
- Route avoids heavily populated areas to minimize potential problems from spills and to reduce opportunities for tampering.
- Route can service large users, such as airfields and logistical support areas.
- Route follows natural linear features, such as wood edges and fencerows to aid in camouflage.

- Route takes into account applicable environmental considerations.
- Route takes into account AT and other protection considerations.

15-21. It is essential to determine elevations along the route as part of the reconnaissance. This data is critical to the system's hydraulic design. The hydraulic design determines the location and number of pump stations and of certain control devices required for the pipeline to work properly.

CONSTRUCTION PHASE

15-22. Different parts of the pipeline system can be built simultaneously. As construction crews are clearing the pipeline route, other crews can be building gap-crossing structures or installing pump stations and intermediate storage facilities. The construction of a pipeline system requires maximum flexibility and decentralized control of the construction elements. Leaders of small units must be well prepared to function with a minimum of supervision because the project will likely have elements spread over many miles. It is because of this that most projects of this scope will have a brigade or battalion level C2 element. In this way, the entire organization can be most effectively employed.

15-23. The tactical situation, terrain difficulty, and required supporting construction will determine how the construction will be carried out. The joining of pipeline elements is likely to be a short end phase, with longer, earlier phases in which the companies work in a decentralized fashion.

15-24. As the pipeline is assembled, certain sections must be pressure-tested carefully to ensure that they are leakproof. Fuel lines need to be equipped with vacuum, pressure, and vapor releases as required. Any section of pipe that cannot be visually inspected or is not readily accessible must meet this criterion. Sections of pipe that are buried underground or are submerged underwater must be tested. Other critical sections include any parts of a pipeline that are placed in tunnels used by personnel or vehicles. Leaks in tunnels may allow vapors to accumulate or expose the pipe to damage from moving vehicles; a fire or explosion may result.

15-25. The pipeline can be checked by pressure-testing with four fluids: potable water, nonpotable water, saltwater, and fuel. The preferred method is to use potable water, as it prohibits corrosion. The engineer unit is responsible for providing water for this event. Water is introduced into the pipeline and subjected to increasing pressure for a period of time. The pipeline must maintain the required pressure for the specified period before the operating unit (the PPTO battalion) will accept that specific section of pipeline. Testing with air can also be used for shorter sections of line, but leaks are difficult to pinpoint. Under extreme operational requirements, fuel may be authorized for testing, but only as a last resort. Applicable environmental considerations should be followed.

PIPELINE MAINTENANCE

15-26. Once the PPTO battalion accepts the pipeline, it is responsible for operational maintenance. The unit will make frequent inspections of the line for visual signs of leaks and damage. The unit is capable of repairing minor leaks and replacing short sections of pipeline that have been damaged. The operating unit will need engineer support to make repairs beyond its capabilities (minor repairs). This includes, but is not limited to, buried pipe or pipe that is in an inaccessible location. Generally, engineers are needed for major repairs, such as 1-mile or more sections of leaking or destroyed pipe that requires reconstruction, or sections that have severe petroleum leaks.

15-27. Safety is extremely important when dealing with pipeline breaks and leaks. Spilled fuel must be contained to reduce the fire hazard and to prevent the contamination of water supplies. Absolute control of all flame- or spark-generating equipment or material within or near the work area is vital. Security measures must be considered against sabotage and pilfering.

Chapter 16

Water Supply and Well Drilling

I don't know what the hell this "logistics" is that Marshall is always talking about, but I want some of it.

Fleet Admiral Ernest J. King, 1942

Maintaining a constant supply of water is critical to sustaining the force. Beyond Soldier consumption, it is critical for CBRN decontamination, sanitation, construction, medical operations, and equipment maintenance. The quantity and quality required depends on climatic conditions, terrain, and the type of operation conducted. Logistics officers at all levels estimate their unit needs for water and plan all required procurement and distribution activities to meet those needs. When part of their plan includes well drilling or support for the establishment of distribution systems, commanders will request GE support.

FIELD WATER SUPPLY

16-1. Tactical and logistical planners determine and coordinate water support functions in the AO. Water distribution units are responsible for distribution. Preventive medicine personnel analyze, test, and certify water supplies. Together, they ensure that there is enough quality water production and distribution to continuously support the forces. Logistics planners, using FM 10-52, estimate the required quantity and quality based on the mission, size of the supported force, dispersion of forces in the AO, and availability of various sources of water supply.

16-2. Logisticians use FM 10-52-1 to plan the actual distribution of water to units. Logistics units, typically supply companies, normally have a water distribution platoon assigned to conduct water supply operations, including establishing and operating a water supply point. To do this effectively, the logistics unit may require engineer support, to include—

- Building combat roads and trails to establish traffic control patterns at the distribution site.
- Constructing improvised dams for impounding small streams to obtain a steady source of water.
- Constructing gravel pads to ensure a steady platform for operating reverse osmosis water purification units (ROWPUs).
- Constructing a brine pit (ROWPU support).
- Digging intake galleries along banks of streams.
- Improving drainage at the facility to prevent muddy conditions that may cause the area to become unusable.
- Constructing pads for water storage blivets.
- Constructing or repairing troop bed-down, protection, AT, and maintenance facilities (because water distribution points are often long-term operations).
- Providing diving support for the emplacement of offshore water hoses and pipelines.
- Rehabilitating damaged wells and distribution points.

16-3. Most water supply units are equipped with two 10-mile segments of the tactical water distribution system. The tactical water distribution system is used to transport potable water from wells, desalination plants, and other sources over distances less than 10 miles (per segment) to 20,000-gallon fabric storage tanks. The system is capable of transporting water forward up to 80 miles at a rate of 600 gallons per minute across level terrain. Engineers support this system by providing GE support to set up the distribution point, leveling water storage pads, and assisting with the emplacement of the hose system. The need to cross hard-surfaced roads and other obstacles will require engineers to install culverts or emplace suspension kits at various locations throughout the hose system.

16-4. GE support in the form of well drilling is provided to the water collection and distribution process. Well drilling may occur when—

- Surface sources of water are not available in enough quantity or quality to support the force. This is likely to occur in arid terrain where the quantity of water required is high and surface sources are low.
- The distribution system is insufficient to support the force. Haul distances may be significantly reduced by a well drilled close to the consumer.
- CBRN or other contamination is expected that would render surface sources unusable.
- The mission is part of a HCA mission. A major portion of the world's population does not have a readily available source of potable water. Providing a source by conducting well-drilling operations may be the decisive operation in a stability and reconstruction operation, and a critical part of the overall IO.

WATER DETECTION

16-5. The detection of groundwater sources is critical for successful well-drilling operations. Without proper analysis, the potential for finding an adequate source is less likely. Determining the most suitable sites to drill for groundwater falls primarily on geospatial teams and the water detection response team (WDRT). Geospatial teams use data from terrain and other geospatial products to recommend the best sites to conduct well-drilling operations. These teams use the results of field reconnaissance and geophysical surveys to provide recommendations. They also have the capability of reachback to experts at the Topographic Engineering Center to obtain data and analysis from historic records and further SME analysis to identify areas with a high potential for developing water supply sources. Geospatial teams are not equipped or trained for actual detection, only predictive analysis.

16-6. The mission of the WDRT is to assist the military planner, terrain team, and all DOD well-drilling teams in locating adequate groundwater supplies before drilling to improve military well-drilling success. In unfamiliar terrain, drilling by trial and error can be costly and time-consuming. When needed, contact the Topographic Engineering Center (see appendix B).

16-7. Well-drilling teams may drill exploratory or test holes to detect groundwater, but this method is time-consuming. It is only recommended if other water detection methods are not available or have been proven to be unsuccessful.

WELL-DRILLING OPERATIONS

16-8. Wells provide water to the deployed forces. Proper planning and execution guidance for the conduct of well-drilling operations resides in FM 5-484/NAVFAC P-1065/AFMAN 32-1072 (note that the Navy and Air Force maintain well-drilling capabilities that are addressed in this manual). Wells should be drilled of a secure area within the AO and, if possible, within base camps or other facilities for which the water will be used. The well-drilling team is inherently modular and deploys to the AO with the organic equipment they use to drill and complete a well.

16-9. Well-drilling teams are a theater engineer command asset and should be deployed and employed by an engineer brigade or battalion capable of providing expertise and logistical support. Since the team has limited personnel, the engineer headquarters must also plan for security at any work site.

16-10. Well-drilling projects should be managed as any construction project. The well-drilling team commander must coordinate work closely with the construction or operations officer of the higher headquarters unit to ensure timely reporting. Because well-drilling equipment is inherently large and heavy, engineers must ensure its mobility in conditions with poor trafficability. As the team moves from project to project, the operations officer arranges the transfer of all completed work to the user, movement to the new project site, and further logistics and security arrangements.

16-11. Drilling rigs are either truck- or semitrailer-mounted. Current well-drilling and well completion equipment consists of the 600-foot, well-drilling system. The system includes—

- A truck-mounted drilling machine mounted on a Navistar™ 6-by-6 truck chassis.
- A truck-mounted tender vehicle.
- A lightweight, well completion kit (including accessories, supplies, and tools needed for drilling a well).

16-12. The Army also uses the CF-15-S trailer-mounted, 1,500-foot, well-drilling machine and 1,500-foot completion kit. The 600-foot well-drilling system has replaced the CF-15-S trailer-mounted machine and 1,500-foot, well completion kit, but the CF-15-S may still be found forward-deployed for contingency operations.

16-13. The 600-foot, well-drilling system (figure 16-1) can be deployed with minimal preparation and support equipment anywhere in the world. With the completion kit, drillers can complete a well to a depth of 600 feet using mud, air, or a down-hole hammer—with or without foam injection. With the augmentation of an auxiliary 250-CFM air compressor, drill pipe, and 400 feet of drilling stem, the 600-foot well-drilling system can drill up to depths of 1,000 feet in a variety of soil conditions using mud or drilling foam. Additional equipment includes casing elevators and slips, larger drill bits, and an additional drill stem. Well-drilling teams should ensure that they have the rig accessory kit for the LP-12 to be fully mission-capable. The 600-foot, well-drilling system is—

- Air-transportable by a C-130, C-141, and C-5. The vehicle is equipped for tie-down as well as lift operations during transport.
- Equipped for air percussion drilling and for rotary drilling with mud or air.
- Equipped to drill wells up to 600 feet.
- Adaptable for drilling to a depth of 1,500 feet.
- Truck-mounted for mobility.
- A three-mode, water transfer pumping system.

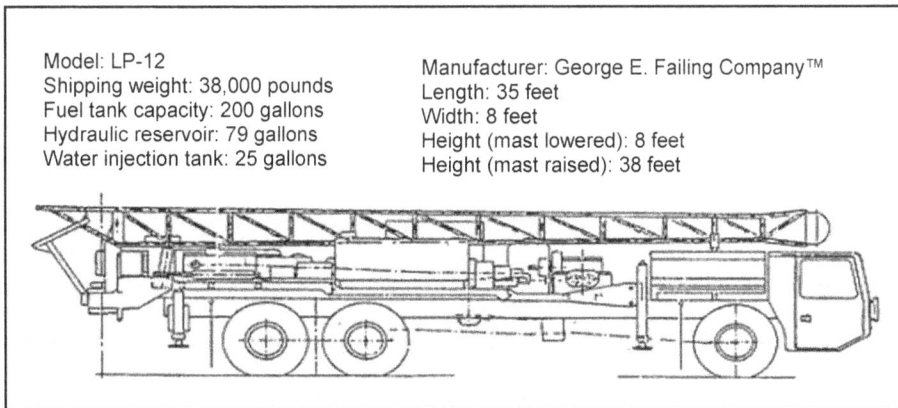

Model: LP-12
Shipping weight: 38,000 pounds
Fuel tank capacity: 200 gallons
Hydraulic reservoir: 79 gallons
Water injection tank: 25 gallons

Manufacturer: George E. Failing Company™
Length: 35 feet
Width: 8 feet
Height (mast lowered): 8 feet
Height (mast raised): 38 feet

Figure 16-1. 600-foot, well-drilling system and specifications

This page intentionally left blank.

Appendix A

Metric Conversion Table

A-1. When planning GE missions, it is often necessary to use metric units to standardize project measurements. Table A-1 is intended to serve as a basic conversion table for that purpose.

Table A-1. Metric conversion table

U.S. Units	Multiplied By	Equals Metric Units
Acres	0.4947	Hectares
Acres	43,560	Square feet
Acres	4,047	Square meters
Bulk fuel, 55-gal drum (bbl)	0.17	Soft tons
Cubic feet	0.0283	Cubic meters
Cubic inches	16.3872	Cubic centimeters
Cubic inches	0.0164	Liters
Cubic yards	0.7646	Cubic meters
Degrees Fahrenheit	Subtract 32, multiply by 5/9	Degrees Celsius
Feet	0.3048	Meters
Feet per second	18.288	Meters per second
Fluid ounces	29.573	Milliliters
Gallons	0.1337	Cubic feet
Gallons	0.00378	Cubic meters
Gallons	3.7854	Liters
Gallons (bulk fuel)	0.004	Soft tons
Inches	2.54	Centimeters
Inches	0.0254	Meters
Inches	25.4001	Millimeters
Miles (nautical)	1.85320	Kilometers
Miles (statute)	1.6093	Kilometers
Ounces	28.349	Grams
Pounds	453.59	Grams
Pounds	0.4536	Kilograms
Square inches	6.4516	Square centimeters
Square feet	0.0929	Square meters
Square miles	2.59	Square kilometers
Square yards	0.8361	Square meters
Yards	0.914	Meters

Table A-1. Metric conversion table

Metric Units	Multiplied By	U.S. Units
Centimeters	0.3937	Inches
Cubic centimeters	0.061	Cubic inches
Cubic meters	35.3144	Cubic feet
Cubic meters	1.3079	Cubic yards
Degrees Celsius	By 9/5 and add 32	Degrees Fahrenheit
Milliliters	0.03380	Fluid ounces
Grams	0.03527	Ounces
Kilograms	2.2046	Pounds
Kilometers	0.5396	Miles (nautical)
Kilometers	0.62137	Miles (statute)
Meters	3.2808	Feet
Meters	39.37	Inches
Meters	1.0936	Yards
Millimeters	0.03937	Inches
Square centimeters	0.155	Square inches
Square kilometers	0.3861	Square miles
Square meters	1.196	Square yards
Square meters	10.764	Square feet

Appendix B
Reachback Tools

The availability of military and civilian engineers through reachback provides the full expertise of the Regiment to support full spectrum operations, enhancing the capabilities and expertise of forward-deployed forces while minimizing the required footprint. This appendix is designed to highlight the most useful and primary support available to engineers performing GE operations. These tools include resources from the USAES, USACE, and the U.S. Army Center for Health Promotion and Preventive Medicine (USACHPPM). This type of reachback capability is one of the characteristics of FFE. The Air Force and Navy provide some of the same type of capabilities and support through the Air Force Civil Engineering Support Agency and the NAVFAC. See FM 3-34 for further information.

UNITED STATES ARMY ENGINEER SCHOOL

B-1. The Doctrine Development Division of the USAES manages engineer doctrine within the TRADOC doctrine development cycle. As part of this process, the Doctrine Development Division assesses, plans, develops, produces, and disseminates engineer doctrine that is synchronized with allied, multinational, joint, multi-Service, and combined arms doctrine. It supports the development of nonengineer doctrinal products by providing subject matter expertise for review and coordination, supports the Engineer Regiment by managing the Center for Engineer Lessons Learned, and provides information and analysis as needed.

UNITED STATES ARMY CORPS OF ENGINEERS

B-2. The USACE has over 35,000 military and civilian employees in 9 USACE divisions, 45 districts, multiple centers of expertise, and laboratories. The military and civilian engineers, scientists, and other specialists work hand in hand as leaders in engineering and environmental matters. The diverse workforce of biologists, engineers, geologists, hydrologists, natural-resource managers, and other professionals meets the demands of changing times and requirements as a vital part of America's Army (see figure B-1, page B-2).

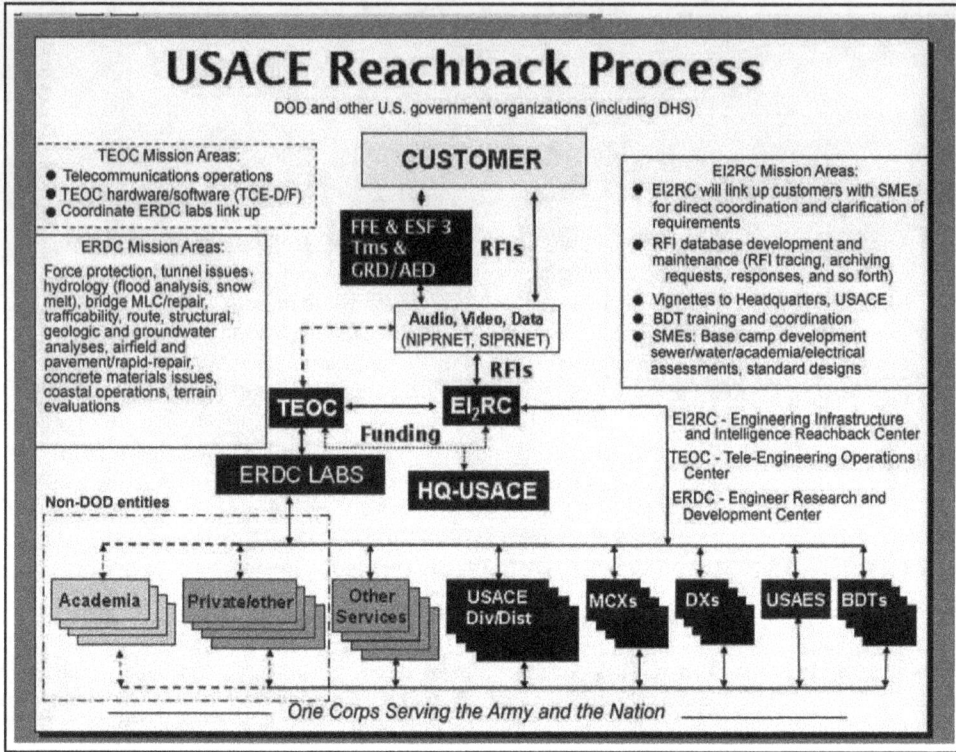

Figure B-1. The USACE reachback process

ENGINEERING INFRASTRUCTURE AND INTELLIGENCE REACHBACK CENTER

B-3. The EI2RC serves as the USACE FFE "hub" for engineering support and GIS infrastructure intelligence to military deployments and civil-military operations worldwide. As such, it is the primary USACE reachback center for technical assistance and engineering support.

B-4. The concept of operations for the EI2RC is as follows:

- The requestor or customer sends a request for information (RFI) to the EI2RC via SIPRNET or Nonsecure Internet Protocol Router Network (NIPRNET) EI2RC Web site, video teleconferencing, telephone, or SIPRNET or NIPRNET e-mail.
- EI2RC accepts RFIs for a variety of technical and nontechnical assistance, including, but not limited to, structural assessment; mapping; satellite imagery; GIS; evacuee, EPW, and base camp development; water resources; protection; road, airfield, and port repair; and expedient designs.
- EI2RC assigns the RFI to a BDT, laboratory, or center of expertise and oversees the completion of the RFI.
- EI2RC archives RFI deliverable products into an easy-to-search, online database.

B-5. The EI2RC is currently fielding the "it knows everything™(ike)" with the geospatial assessment tool for engineering reachback (GATER), a field data collection platform for collecting information related to infrastructure assessments in the field. The ike is a rapid data collection device consisting of a handheld computer, global positioning system (GPS), laser range finder, digital camera, and inclinometer all in one ruggedized handheld device. The GATER is actually the customized ArcPAD® interface that is developed and tailored for your specific AOR. The GATER currently supports infrastructure assessments as a module of forms that the user easily accesses and fills in during the field infrastructure assessment. The deployment of the ike device with the GATER is considered ike with GATER (figure B-2 and figure B-3).

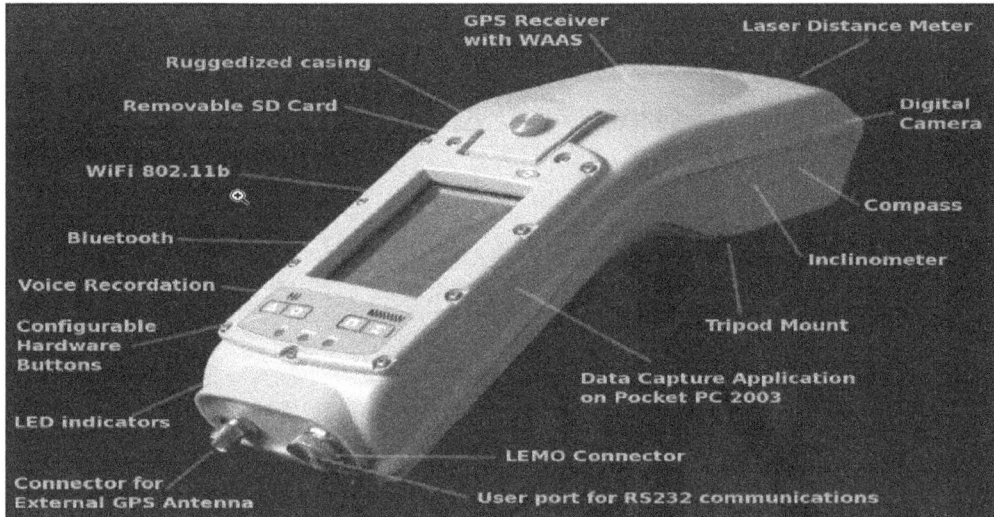

Figure B-2. Ike

Figure B-3. GATER

B-6. The EI2RC also has the repository and archiving capability to provide Reachback technical assistance to military deployments and civil-military operations. The EI2RC is designated as the repository for the information, document, and designs it generates and is therefore responsible for archiving and maintaining an online NIPRNET and SIPRNET document management system. This Web-based system provides the means to receive, locate (spatially), assign, track, and archive all documents from RFI deliverables. Also available are the situation reports (SITREPs), after-action reviews, lessons learned, standard designs, and technical library documents. The online system has been customized so that supported units can easily submit their requests and later query the document management system to download response deliverables.

TELE-ENGINEERING OPERATIONS CENTER

DESCRIPTION

B-7. The TEOC provides a reachback engineering capability that allows DOD personnel deployed worldwide to talk directly with experts in the United States when a problem in the field needs quick resolution. Deployed troops can be linked to SMEs within the USACE (or comparable Air Force and Navy organizations), private industry, and academia to obtain detailed analysis of complex problems that would be difficult to achieve with the limited expertise or computational capabilities available in the field.

CAPABILITIES

B-8. TEOC staff members respond to incoming information requests and provide detailed analyses of problems that the ERDC laboratories can address, such as flooding potential due to dam breaches, load-carrying capacities of roads and bridges, field fortifications and protection, evaluation of transportation networks, and water resource data.

SUPPORTING TECHNOLOGY

B-9. TCE-D provides reachback capability using COTS communications equipment with encryption added. Video teleconferences and data transfers can be conducted from remote sites where other similar means of communications are nonexistent or unavailable.

B-10. The tele-engineering toolkit (TETK) is a software product that provides a valuable analysis tool to personnel on the ground or going into an AO. By annotating an area of interest, a small reference file can be sent back to the SMEs to provide requests for a variety of information, to include cross-country mobility analysis, flood analysis, and vegetation information. The response can then be sent back and graphically displayed using the TETK.

ENGINEER RESEARCH AND DEVELOPMENT CENTER

B-11. The U.S. Army ERDC is one of the most diverse engineering and scientific research organizations in the world. It consists of seven laboratories at four geographical sites in Vicksburg, Mississippi; Champaign, Illinois; Hanover, New Hampshire; and Alexandria, Virginia. It employs more than 2,000 engineers, scientists, and support personnel.

B-12. The ERDC supports the DOD, other federal agencies, and the nation in military and civilian projects. Its primary mission areas include the following:
- Warfighter support.
- Installations.
- Environment.
- Water resources.
- Information technology.

B-13. Research projects include facilities, airfields, pavements, protective structures, sustainment engineering, environmental quality, installation restoration (cleanup), compliance, conservation, regulatory functions, flood control, navigation, recreation, hydropower, topography, mapping, geospatial data, winter climatic conditions, oceanography, environmental impacts, and information technology.

UNITED STATES ARMY CORPS OF ENGINEERS CENTERS OF EXPERTISE AND LABORATORIES

B-14. The USACE has numerous centers of expertise and laboratories who conduct research and development for both military and civil missions the Corps supports. Listed below is the majority of centers of expertise, directed centers of expertise (DXs), and laboratories within USACE.

ENGINEERING RESEARCH AND DEVELOPMENT CENTER

B-15. The following is a list of the ERDCs:
- Geotechnical and Structures Laboratory, Vicksburg, Mississippi.
- Coastal and Hydraulics Laboratory, Vicksburg, Mississippi.
- Environmental Laboratory, Vicksburg, Mississippi.
- Information Technology Laboratory, Vicksburg, Mississippi.

- Topographic Engineering Center, Alexandria, Virginia (Fort Belvoir).
- Construction Engineering Research Laboratory, Champaign, Illinois.
- Cold Regions Research and Engineering Laboratory, Hanover, New Hampshire.
- Computer-Aided Design and Drafting (CADD) and GIS Technology Center for Facilities, Infrastructure, and Environment, Vicksburg, Mississippi, DX.
- Concrete Technology Information Analysis Center, Vicksburg, Mississippi, DX.

UNITED STATES ARMY ENGINEERING AND SUPPORT CENTER, HUNTSVILLE, ALABAMA

B-16. The following are located at the United States Army Engineering and Support Center:

- Ballistic Missile Defense.
- Chemical Demilitarization.
- Installation Support.
- Medical Facilities Center of Expertise.
- Ordnance and Explosives.
- Ranges and Training Land Program.
- TCMS.
- Electronic Security Systems.
- Electronic Technology Systems Center.
- Utility Monitoring and Control Systems.
- Energy Savings Performance Contracting.
- Facility Repair and Renewal.
- Operation and Maintenance Engineering Enhancement (OMEE).
- Utility Systems Privatization.
- Contingency Facilities Support.
- Facility Standards and Criteria.
- Facility Programming and Cost Engineering and Estimating Tools.

MISCELLANEOUS

B-17. The following are additional DX locations:

- Hydroelectric Design Center, Portland, Oregon, DX.
- Management and Curation of Archeological Collections, Saint Louis, Missouri, DX.
- Preservation of Historic Buildings and Structures, Saint Louis, Missouri, DX.
- Photogrammetric Mapping, Kansas City, Kansas, DX.
- Heating, Ventilation, and Air Conditioning Control Systems, Savannah, Georgia, DX.
- Automated Performance Monitoring of Dams, Saint Louis, Missouri, DX.
- Remote Sensing/GIS Center, Hanover, New Hampshire, DX.
- Hydrologic Engineering Center, Davis, California, DX.
- Subsurface Exploration Center, Mobile, Alabama, DX.
- Readiness Support Center, Mobile, Alabama, DX.
- Internet Center of Expertise, Hanover, New Hampshire, DX.
- National Security Agency Real Estate Technical Support Center, Baltimore, Maryland, DX.
- Technical Assistance Center (Real Estate) Technical Support Center, Savannah, Georgia, DX.
- U. S. Army, Southern Command Real Estate Technical Support Center, Mobile, Alabama, DX.
- Real Estate Systems Support Center, Mobile, Alabama, DX.
- Defense National Relocation Program, Baltimore, Maryland, DX.
- Aircraft Hangar Fire Protection, Winchester, Virginia, DX.
- Marine Design Center, Philadelphia, Pennsylvania.

- Protective Design Center, Omaha, Nebraska.
- Transportation Systems Center, Omaha, Nebraska.
- POL Fuels, Omaha, Nebraska, DX.
- Hazardous, Toxic, and Radioactive Waste, Omaha, Nebraska.
- Rapid-Response Hazardous, Toxic, and Radioactive Waste Center, Offutt Air Force Base, Nebraska, DX.
- Photogrammetric Mapping, Saint Louis, Missouri, DX.
- Construction Equipment/Cost Index Directory of Expertise, U.S. Army Corps of Engineers, Walla Walla, Washington.
- Wastewater Treatment, Mobile, Alabama, DX.
- Power Reliability Enhancement Program, Headquarters USACE, Fort Belvoir, Virginia.

THEATER CONSTRUCTION MANAGEMENT SYSTEM

DESCRIPTION AND PURPOSE

B-18. The TCMS is a personal, computer-based automated construction planning, design, management, and reporting system that is used by military engineers for contingency construction activities. Its primary purpose is to support engineer planners with facilities design information for outside the continental United States (OCONUS) and contingency mission requirements. It is intended to be used by all levels of engineer units from the theater engineer command down to the engineer company level.

B-19. The TCMS is the approved method for distributing the AFCS designs and related information. The AFCS provides logistical and engineering data that is organized, coded, and published to assist engineer planners and designers in executing GE missions in contingency environments. AFCS is governed by the following:

- AR 415-16.
- TM 5-301-1.
- TM 5-301-2.
- TM 5-301-3.
- TM 5-301-4.
- TM 5-302-2.
- TM 5-302-3.
- TM 5-302-4.
- TM 5-302-5.
- TM 5-303.

B-20. The system determines personnel and materiel requirements and the cost, weight, and volume of materials needed for a specific project. It provides standard plans for base camp development, utilities, and airfields. The TCMS is updated and distributed annually.

B-21. The TCMS provides the automation tools necessary to use the AFCS designs information to accomplish TO engineering and construction activities in support of mission requirements. It contains detailed project descriptions and related construction estimates to include the following:

- Real estate requirements.
- Design and construction drawings and plans.
- BOM for individual facilities or the complete project.
- Construction resource estimates as related to Army engineer unit construction capability (will be updated to include other Service engineer unit capabilities as well).
- Theater-oriented construction guide specifications.
- Construction directives.

- DD Form 1391 (FY ____ Military Construction Project Data) process initiation.
- Project and unit construction status reports.

THEATER CONSTRUCTION MANAGEMENT SYSTEM ASSESSMENT

B-22. Distribution of TCMS is available upon request to all U.S. military engineer units, including all Active Component, USAR, and ARNG. This includes requests from all other Service engineers.

B-23. The Huntsville Center of the USACE provides active TCMS support and trains TCMS users in the basic operation of the system. To take full advantage of the system, users will need to know how to use commercial software packages. Automated computer-assisted design (AutoCAD®) and Microsoft Project™ constitute a large portion of the TCMS capability.

KEY FEATURES OF THEATER CONSTRUCTION MANAGEMENT SYSTEM

B-24. Figure B-4 depicts the TCMS online system.

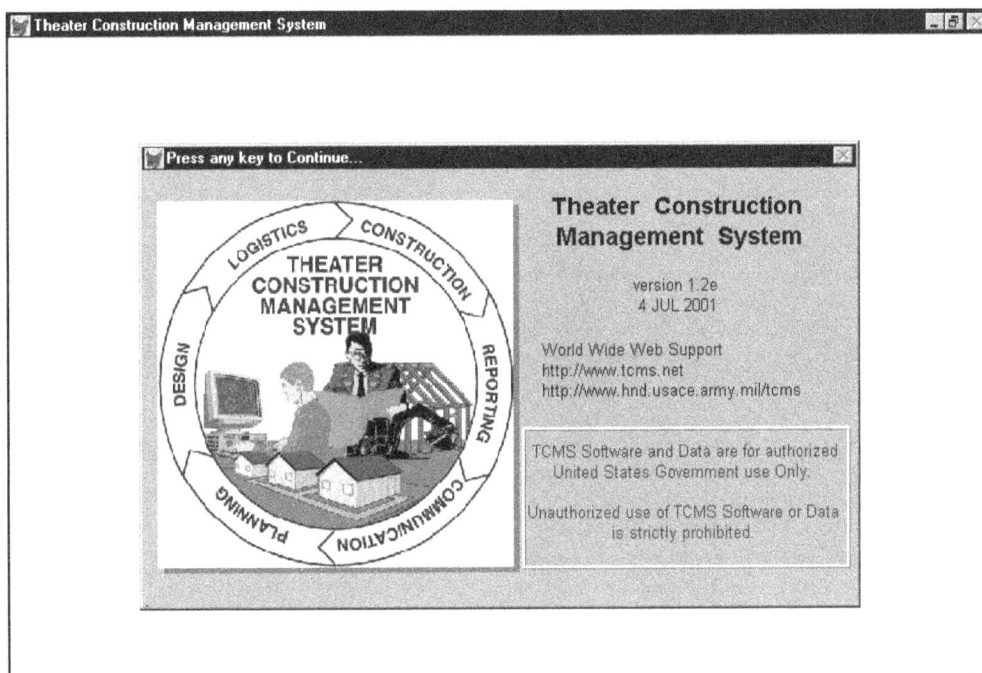

Figure B-4. TCMS online

Planning

B-25. Users are able to develop facility and installation plans to satisfy mission construction requirements using the TCMS computer routines. Planning information includes providing estimates for material and construction requirements.

Design

B-26. Users are able to prepare site-specific and new design and construction drawings, use existing AFCS designs within TCMS, or modify as required to site. They can adapt mission requirements using the TCMS CADD capability.

Management

B-27. Users are able to set up and manage the construction progress and the construction resource allocation and utilization throughout the construction time frame.

Reporting and Communication

B-28. The TCMS develops and transmits the necessary reports up the engineer chain of command. This facilitates the decision-making process using inter computer electronic and direct entry.

WATER RESOURCE DETECTION TEAM

PURPOSE

B-29. The objective of the water resource detection team (WRDT) is to identify high-potential areas for the best quality of water, within available drilling equipment capability, to meet the water production requirements of the mission.

CAPABILITIES

B-30. Water resources database (WRDB) expertise and studies are concentrated in four areas or elements. The four areas or elements are listed below:

- **Database.** The Topographic Engineering Center produces and maintains a worldwide DOD WRDB of available water supply and hydrologic data, including groundwater resources. The WRDB is derived from classified and open-source data, maps, documents, and imagery. When specific missions and requests are received for areas where data are uncertain or inconclusive, the team will research additional sources and data unique to the area. The resulting WRDT product or report summarizes the information critical to planning a successful well, such as the hydrogeology, target depth, aquifer material, expected yield, and probable water quality. Office studies, based on research and analysis of existing data, are the most cost-effective and timely WRDT approach and take hours to days to complete.

- **Remote sensing.** If databases and other supplemental information are inadequate, aerial or satellite imagery may be studied and analyzed for indications of groundwater. This source is especially useful in a hard-rock area, where siting wells on significant fractures and fracture intersections is the key to success. The acquisition and analysis of imagery increases the time and cost to complete an office study.

- **Supporting specialists.** If office studies including imagery analysis are inadequate, one or more supporting specialists may be deployed to the site. These specialists contact HN groundwater experts, collect and evaluate in-country data associated with existing or historic wells, and conduct hydrogeologic field reconnaissance of specific areas before drilling. They may also assist with interpreting well cuttings and down-hole electric logging during drilling. Field studies take days to weeks to complete.

- **Geophysics.** Should information gathered by supporting specialists be insufficient, additional local site investigation may be necessary using exploratory geophysics. Geophysicists may deploy to the site to conduct electrical resistivity, seismic refraction, or other on-site tests to better define the subsurface before drilling. Geophysical exploration and data analysis generally take weeks to complete. Costs are significantly higher than for office-based studies and are paid by the requester.

B-31. When activated, the WRDT does not automatically deploy to provide technical support for military operations. The starting point for each WRDT request is to identify high-potential areas through the examination of existing databases, followed by the collection and analysis of additional sources and imagery. In those rare cases when high-potential water sites cannot be identified from source data and imagery, teams from the supporting specialists and/or geophysics elements can be deployed for on-site investigations. This should take place before the arrival of the well drillers. If deployed to the TO, the WRDT operates as a component of the theater engineer command or senior engineer organization in theater. As with any USACE capability, activation of the WRDT for deployment is not automatic; it must be requested through the supporting theater engineer command or through other appropriate command channels to the Topographic Engineering Center. The commander provides and arranges for the WRDT logistics and administrative support necessary for mission accomplishment.

AUTOMATED ROUTE RECONNAISSANCE KIT

PURPOSE

B-32. An ARRK provides military units an adaptable, easy-to-use reconnaissance package. This allows engineer reconnaissance teams or other engineer reconnaissance elements to rapidly collect and process reconnaissance (all types, but generally route) information.

CAPABILITIES

B-33. The ARRK uses a field-ready laptop computer to continuously collect reconnaissance information without stopping or leaving the vehicle for routine calculations. Time, security, and accuracy issues associated with a route reconnaissance are reduced. The ARRK collects pictures, voice recordings, GPS locations, accelerometer, and gyroscope data streams in three dimensions. Unlike the traditional, manually recorded route reconnaissance efforts, the ARRK allows an operator with minimum training experience to collect, process, and export route information. The ARRK accommodates a chronological, picture replay of the route and georeferenced display of major features that affect the classification and usage of the road or route. The viewer of the data can scroll through the stored data types to instantly locate specific features along the route. Data includes automated determination of slope, radius of curvature, and ride quality. The reconnaissance data collected from the ARRK is quickly converted by the operator to a preformatted report that is according to the requirements of FM 3-34.170. Planned improvements for the system include the integration of a laser range finder and digital scale reference guide. The system has the capability to be a stand-alone data collection tool and a fully interoperable data collection platform for dissemination and a repository of route information.

ANTITERRORISM PLANNER, VERSION 2.1

PURPOSE

B-34. The AT Planner is based on threat, mission, and site considerations. It is designed to assist commanders in planning, implementing, and evaluating protective measures, expedient structure designs, and standoff guidance required for protection.

CAPABILITIES

B-35. The AT Planner provides the user with a computerized analysis tool for evaluating critical assets in terrorist threat scenarios based on aggressors, tactics, and weapons systems. The threat conditions dictate a number of security measures from this manual that the user must consider and possibly employ. These measures are cumulative from the lowest to the highest threat level and are presented in the AT Planner in a concise format. Emphasis has been placed on the evaluation of structural components, windows, personnel, and other limited critical assets. Structural components are defined as frames, walls, and roofs from common construction materials. Damage to the building components is calculated using algorithms from the Facility and Component Explosive Damage Assessment Program (FACEDAP), with the user providing the distance of the explosive charge from the building.

B-36. The AT Planner can provide the required standoff for a given explosive charge. Once the appropriate standoff is determined based on expected explosive size and an acceptable level of building damage, the AT Planner provides information on protective barriers and a vehicle velocity calculator to aid in barrier and obstacle selection. Extensive information is available on various types of obstacles and protective barriers and the information source is referenced. In addition, the AT Planner provides a basis for the design and analysis of wall and window retrofits. The capability is available to view facility or site images, locate assets on the site image, and show building damage in 2-D and 3-D graphical formats. Blast walls can be placed in front of structures, and the resulting damage to a protected building is then calculated. Glass hazard calculations have been incorporated along with user-defined, pressure-impulse curves to give engineers more flexibility in evaluating structures.

B-37. The AT Planner contains algorithms to estimate injuries and fatalities to occupants of structures and provides the consequences of a terrorist attack for the AT program planning. The AT Planner can be obtained by contacting ERDC.

DEFENSE ENVIRONMENTAL NETWORK AND INFORMATION EXCHANGE

PURPOSE

B-38. The Defense Environmental Network and Information Exchange (DENIX) is an electronic environmental bulletin board accessible throughout the DOD. It gives DOD environmental, safety, and occupational health managers a central communications platform to gain timely access to vital environmental information.

CAPABILITIES

B-39. DENIX, which was based on the Army's Defense Environmental Electronic Bulletin Board System, gives users the ability to—

- Read online environmental publications (proprietary or DOD-specific).
- Send and receive mail electronically on the DENIX host computer or across the Internet.
- Exchange environmental information via managed discussion forums based on a subject area.
- Send and receive required reporting data through the chain of command.
- Peruse and request environmental training courses and seminars.
- Access the DENIX directory service database.
- Upload and download files from DENIX to and from a personal computer.

UNITED STATES ARMY CENTER FOR HEALTH PROMOTION AND PREVENTIVE MEDICINE

PURPOSE

B-40. The USACHPPM has a mission to provide worldwide technical support for implementing preventive medicine, public health, and health promotion and wellness services into all aspects of America's Army and the Army community, anticipating and rapidly responding to operational needs and adaptable to a changing world environment. The USACHPPM organization is headquartered at Aberdeen Proving Ground, Maryland, with subordinate commands in Landstuhl, Germany, and Sagami, Japan. USACHPPM also maintains subordinate commands at three locations across the United States: Fort Meade, Maryland; Fort McPherson, Georgia; and Fort Lewis, Washington.

CAPABILITIES

B-41. The USACHPPM is a linchpin of medical support to combat forces and the military managed-care system. It provides worldwide scientific expertise and services in clinical and field preventive medicine, environmental and occupational health, health promotion and wellness, epidemiology and disease surveillance, toxicology, and related laboratory sciences. It supports readiness by keeping Soldiers fit to fight, while also promoting wellness among their families and the federal civilian workforce. Professional disciplines represented include chemists, physicists, engineers, physicians, optometrists, epidemiologists, audiologists, nurses, industrial hygienists, toxicologists, entomologists, and many others as well as subspecialties within these professions.

Appendix C

Infrastructure Rating

Table C-1, pages C-1 through C-4, is provided as an example of how the staff engineer (as the integrating staff proponent) can communicate the results of an infrastructure assessment or survey to the commander. While the areas are generally the same as the areas included in the infrastructure assessment worksheet, the evaluation criteria for some of the rated areas may require more specific information, such as that provided in the survey. It is intended that the individual conducting the assessment or survey also complete the infrastructure rating as a part of a complete infrastructure reconnaissance. See chapter 4 for more information on infrastructure assessment and survey.

Table C-1. Status color coding of infrastructure categories

Area	Green	Amber	Red	Black
Sewage	Sewage system works consistently	Sewage system works, but treatment status is undetermined	No treatment observed, but treatment plant exists	No sewage treatment system; destroyed
	No sewage observed and no odor	No sewage observed, but odor is present and/or system is damaged	Sewage observed, and odor present	Presence of raw sewage is a public health issue
	Operational in 100 percent of public facilities	Operational in 50 percent or more of public facilities	Operational in less than 50 percent of public facilities	No operational sewage in public facilities
Water	Water distribution works at 100 percent capacity	Water distribution works at 50 percent or more of capacity with some leaks	Water distribution does not work	No water distribution system; destroyed
	Tested as clean and/or local populace is consuming	Appears clean, no smell, and local populace states that it is clean	Does not appear clean, and local populace states that it is not clean	Tested, nonpotable and/or appears contaminated and has bad odor
	Running water in 100 percent of public facilities	Running water in 50 percent or more of public facilities	Running water in less than 50 percent of public facilities	No running water in public facilities
Electricity	Power distribution system works; blackouts are planned	Power distribution system works; blackouts unplanned	Power distribution system is unreliable; frequent blackouts	No power distribution system; destroyed
	Electric lines are 100 percent; no damage and no power loss	Electric lines are 50 percent; some minor damage and undetermined power loss	Electric lines are less than 50 percent; major damage and noticeable power loss	Electric lines are all down; hot wires exposed; significant power loss
	Power grid station intact; secure	Power grid station operational; unsecure	Power grid station nonoperational; unable to secure	Power grid station stripped; destroyed
Academics	Building serviceable; all utilities operational; secure	Building is adequate; utilities operate over 50 percent; not secure	Building is usable; utilities operate less than 50 percent; not secure	Building is not usable; utilities are nonfunctional
	Academic resources available to all students	Academic resources available to 50 percent or more	Academic resources available to less than 50 percent	Extremely limited academic resources

Table C-1. Status color coding of infrastructure categories (continued)

Area	Green	Amber	Red	Black
Trash	Formal trash collection system is operational	Formal trash collection system exists, but is limited	No formal trash collection system	No trash collection
	Trash collection is in a central area that does not present a health hazard	Unknown central trash collection area	Central trash collection area presents a possible health hazard	Trash is consolidated in an area that presents a health hazard
	No trash buildup in public facilities	Limited trash in public facilities; relatively clean	Public facilities have no means to remove trash	Public facilities have excess trash
Medical	Medical facilities are functional; backup power; minimal equipment issues; secure	Medical facilities are usable; no backup power; some equipment shortages; not secure	Medical facilities are unsanitary; significant equipment and supply shortages	Medical facilities are not usable due to damage and unsanitary conditions; looted
	Emergency services including multiple ambulatory services available	Emergency services exist; ground transport only	No emergency services; ground transport without medically trained personnel	No emergency services
	Veterinary services available; animal holding area	Limited veterinary services available; inadequate holding area	On-call veterinary services; no holding area	No veterinary services
Safety	Police department functional; secure building; equipment available and operational	Police department functional a minimum of 50 percent; building securable; equipment available and operational less than 50 percent	Police department functional less than 50 percent; unable to secure building; limited equipment available	Police department is nonfunctional; building is not usable; no equipment
	Fire department functional; secure building; equipment available and operational	Fire department functional a minimum of 50 percent; building securable; equipment available and operational more than 50 percent	Fire department functional less than 50 percent; unable to secure building; limited equipment available	Fire department is nonfunctional; building is not usable; no equipment
Other Considerations: Roads and Railroads	Minimum of a Class C road; can be upgraded; no visible damage	Minimum of a Class D road; damage and upgrade requirements will impact traffic flow	Minimum of a Class E road; upgrade requirements are significant; materials not readily available	Road is not trafficable
	Operational railroad system	Railroad is damaged, but resources to repair are available; jacks available	Railroad damage is extensive; resources to repair are not readily available	Railroad system did exist, but now has extensive damage to both track and trains
Other Considerations: Bridges and Waterways	Bridges are trafficable; no visible damage	Bridges are trafficable; damage to spans; supports intact	Bridges are not trafficable for military; risky for civilians; damage to spans and supports	Bridge is not trafficable and is impassable
	MLC verified through ERDC or other structural engineer	MLC calculated, but not verified due to damage	MLC is ineffective due to damage	Construction repair required before MLC can be determined
	Inspection and evaluation shows original strength assessment valid	Inspection and evaluation determines strength support issues	Inspection and evaluation determines minimal supportable strength	Inspection and evaluation determines that bridge cannot support weight

Table C-1. Status color coding of infrastructure categories (continued)

Area	Green	Amber	Red	Black
Other Considerations: Airports	Airport capable of supporting military and civilian traffic concurrently; no visible damage	Airport can support limited military traffic; no visible damage	Airport damaged; utilities and structures are not reliable or safe	No working airport
	Runway, taxiway, and parking aprons are serviceable; working and parking MOG greater than or equal to 2 (military)	Runway serviceable, but taxiway and parking limited; C130 and C17 only	Runway is not serviceable; can repair with available resources	Runway is not serviceable; dimensions will not support military aircraft; major repair and upgrades required
Other Considerations: Housing	Residences are structurally sound and offer protection from the environment	Residences are damaged and need structural evaluation; offer limited protection from the environment	Residences are damaged and structurally unsafe; no protection from the environment	Residences are destroyed
	Utilities are working and reliable	Utilities are working over 50 percent; not reliable	Utilities work less than 50 percent; require significant repairs	Utilities are nonoperational
Other Considerations: Communications	Telephone system operational and reliable in public facilities	Telephone hookups available; some equipment available; somewhat reliable	Limited telephone hookups and equipment available; not reliable	No telephone hook-ups or equipment
	Postal system is operational and reliable	Postal system is slow; over 50 percent of mail delivered	Postal system exists; extremely slow; less than 50 percent of mail delivered	No postal system
	Media—television, internet, radio, newspaper—operational, available, and reliable	One form of media exists and is operational, available, and reliable	One form of media exists but has limited availability and reliability	No form of media
Other Considerations: HAZMAT	HAZMAT/hazardous waste properly segregated, stored, and labeled	Some HAZMAT and hazardous waste not properly segregated, stored, or labeled	HAZMAT and hazardous waste not properly segregated, stored, or labeled	HAZMAT and hazardous waste not segregated, stored, or labeled
	Containers adequate for the material	Containers not generally adequate, but limited corrosion or damage	Containers inadequate, corroded, and leaking	Containers inadequate, corroded, and leaking
	Safety measures and secondary containment in-place	Inadequate safety measures and secondary containment	No safety measures or secondary containment	No safety measures or secondary containment
	Hazards communications system in place	Limited hazards communications system	No hazards communications system	No hazards communications system
	No leaks or spills	Potential for leaks and spills	Some leaks and spills already present; contaminants may enter air, soil, groundwater, or water courses	Gross contamination present; contaminants have entered air, soil, groundwater, and water courses
	Spill prevention and cleanup measures in place/available	Limited spill prevention and cleanup measures available	No ability to prevent or clean up spills	No ability to prevent or clean up spills

Table C-1. Status color coding of infrastructure categories (continued)

Area	Green	Amber	Red	Black
Other Considerations: Attitude	Community leaders not hostile; religious centers are intact; supportive of GE effort	Community leaders are neutral; religious centers are damaged but securable	Community leaders are negative; religious centers are damaged and not securable; skeptic of GE support	Community leaders hostile; religious centers destroyed; do not want GE assistance
	No ethnic tension	Distinct ethnic groups within AO; supportive of GE effort if equal among groups	Distinct ethnic groups within AO; one group dominant; GE tasks cannot be accomplished for all groups	Ethnic violence occurs; one group extremely dominant; GE effort would increase ethnic tension
	Unemployment is less than 50 percent	Unemployment is greater than 50 percent; Willing and able to work to support GE effort	Unemployment is greater than 50 percent; Unable to support GE work effort	Unemployment is a serious issue; unwilling to support GE work effort
	No formal paramilitary threat	Paramilitary threat briefed at the BCT/RCT level	Paramilitary threat a concern at BCT and RCT level	Paramilitary threat a concern at echelons above BCT and RCT level
Note. Food supply and cultural, historical, and religious are still under development.				

Appendix D

Environmental Considerations

GE involves a significant number of environmental considerations in the planning and execution of military operations. The level of environmental considerations applied to a particular military operation will be articulated in annex L of the CCDR's OPORD and must be linked to the engineering decisions and guidance articulated in the engineering support plan. The engineer and ENCOORD on the staff is responsible for ensuring that these considerations are appropriately integrated into the OPLAN and followed in the execution of GE tasks and missions. Efforts should always be made to minimize the release of hazardous substances into the environment, protect cultural and natural resources, and prevent pollution. Operational necessity may at times supersede these concerns, but that is never an excuse for not considering them and minimizing detrimental effects whenever possible. Commanders protect the natural and cultural environment in which U.S. forces operate, to the greatest extent possible, consistent with operational requirements. Environmental restoration may be performed by engineer troop units as GE tasks, but this is generally in support of the USACE when the area is being considered for other uses or closure.

GENERAL ENGINEERING AND ENVIRONMENTAL CONSIDERATIONS

D-1. Environmental considerations are not issues for only the engineer, and yet the engineer is the staff proponent for the integration of environmental considerations. This is very similar to the responsibility that the S-2, G-2, or J-2 has for the integration of the IPB. In both cases, the integrator on the staff works with and assimilates all staff efforts in his respective AOR; and in both cases, the respective staff officer has the responsibility for the creation of an annex/appendix within the OPLAN/OPORD. Environmental considerations are factors that the engineer integrates as the engineer staff running estimate is done. Obtaining knowledge of environmentally sensitive areas and associated impacts is done by coordinating with other staff officers like the judge advocate general; surgeon; S-2, G-2, and J-2; public affairs officer; and others. The combatant command and subordinate joint force engineer and staff develop and assist with the implementation of the policies, procedures, and practices of the environmental considerations annex to an OPLAN and/or OPORD. See annex L, Environmental Considerations, to CJCSM 3122.03A for additional information. In Army plans and orders, appendix 2, Environmental Considerations, to annex F, Engineer, is the parallel document to annex L in a joint OPLAN/OPORD. See FM 3-100.4 for additional information.

D-2. Because of the nature of GE and its potentially large impact on the environment, engineers must be very sensitive to environmental considerations in all GE tasks and missions. While the risks associated with close combat may in some situations override those associated with damage to the environment, GE projects typically are able to give more weight to the risks associated with applying environmental considerations. General guidance on this should be provided in the OPLAN of the CCDR and then further linked to and specified in the engineering support plan.

D-3. Evolving guidance on environmental protection may be forthcoming through the establishment of a JEMB. A JEMB may be established by the CCDR or subordinate JFC for a joint operation to integrate the environmental-protection efforts of all participating components under a single authority, ensuring the unity of effort for environmental protection and other environmental consideration activities. The JEMB should be chaired by the combatant command or subordinate joint force (J-4) or joint force engineer, and include representatives from each Service component and joint force staff as necessary (legal, occupational health, preventive medicine, safety, comptroller, planning, operations, and logistics). The JEMB assists the JFC in establishing the joint force environmental policies, practices, procedures, and priorities and in providing oversight of environmental protection standards and compliance. Establishing a dedicated and appropriately staffed environmental engineering staff, supported by experts from other joint force staff members (legal and medical), may eliminate the need for a JEMB in smaller operations.

ENGINEER COORDINATOR ROLES AND RESPONSIBILITIES

D-4. The ENCOORD is the special staff officer for coordinating engineer assets and operations for the command. As the senior engineer officer in the force, he advises the commander on environmental issues, to include the command environmental program. Working with other staff officers, the ENCOORD determines the impact of operations on the environment and integrating environmental considerations into the decision-making process. He works with the G-4 in performing site assessments for installations and facilities. With the SJA, the ENCOORD advises the commander on the necessity for environmental assessments to meet HN or Executive Order (EO) 12114 requirements. He is also responsible for advising the J-2, G-2, and S-2 of significant environmental factors and ensuring that these impacts are integrated into the IPB process. See FM 34-130 for more information. Additionally, FM 5-0 directs that OPLANs, OPORDs, and concept plans (CONPLANs) contain an appendix to address environmental considerations. Through coordination with the unit staff, the ENCOORD prepares appendix 2, Environmental Considerations, to annex F, Engineer. See FM 3-100.4 for the specific appendix format and content.

D-5. Engineers proactively advise the commander on environmental issues which promote awareness in the unit and success during operations. As the commander's lead advisor for integrating environmental considerations and working with other staff officers, the engineer determines the impact of the operation, mission, and task on the environment and integrates environmental considerations into the decision-making process. Equally important is integrating the associated FHP issues of the effect of the environment on personnel, and recognizing how the environmental changes caused by GE and other tasks/missions can ultimately create changes in the environment that then have a corresponding environmental effect on personnel. Cause and effect relationships must be well understood.

D-6. Commanders, the engineer, and other staff officers should refer to AR 40-5, DA Pamphlet 40-11, FM 3-34, FM 3-100.4, FM 5-0, FM 6-0, Technical Bulletin (TB) Medical (MED) 577, and Training Guide (TG) 248 for guidance in applying appropriate environmental consideration procedures during military operations.

D-7. The integration of medical expertise is especially critical. The combatant command and subordinate joint force surgeon are responsible for health services support (preventive medicine and occupational health) to the joint force. Priorities are on water and wastewater, including water vulnerability assessment support, sanitation, waste disposal (hazardous and infectious waste); health risk assessment (base camp site selection); environmental health sampling; and surveillance and vector control to protect human health and welfare.

ENVIRONMENTAL PLANNING

GENERAL REQUIREMENTS

D-8. Environmental requirements can be divided into overseas requirements and requirements applicable in the United States and its territories and possessions, although some U.S. environmental requirements may have extraterritorial application. For example, EO 12114 establishes requirements for the conduct of environmental studies for activities conducted overseas, somewhat similar to the environmental analysis requirements mandated by the National Environmental Policy Act regarding operations conducted within the United States. The SJA should be consulted to determine extraterritorial applicability. The activation and incorporation of environmental management systems is critical for all DOD organizations (see EO 13148), regardless of whether they operated within the United States or outside of the United States and its territories. Requirements applicable within CONUS may not be applicable to OCONUS situations that are linked to applicable international treaties, conventions, overseas environmental baseline guidance documents, SOFAs, field governing standards, and other international agreements. Finally, in some contingency situations, formal standards and requirements may not exist and it will be necessary for the CCDR to articulate standards and requirements in annex L of the appropriate OPLAN or OPORD. See FM 3-100.4 for a more focused discussion on environmental requirements affecting Army operations and, more specifically, contingency or expeditionary operations.

ENVIRONMENTAL PLANNING NEEDS

D-9. By considering environmental issues early during the planning process, the commander may continue to achieve operational objectives while minimizing the impact on human health and the environment. Failure to consider the environmental impact of all activities may adversely affect the operation. Potential effects include endangering personnel, delaying operation commencement, limiting the future use of exercise or HN areas, and creating an adverse public opinion. Through early assessment of environmental considerations, commanders may become aware of the potential environmental effects or impacts of mission accomplishment while alternatives to address mitigating actions still exist. By planning early, the commander and staff must be aware of the environmental requirements and be able to plan more effectively and act to incorporate environmental considerations. Furthermore, careful and visible attention to environmental considerations in military operations can assist in shaping a positive image, both internationally and domestically.

D-10. Integrating environmental considerations into planning is very similar to integrating safety and protection issues. FM 3-100.4 discusses environmental planning and focuses on how and where the Army integrates environmental considerations into the MDMP. As part of the MDMP, risk management is an effective process to minimize actions that may negatively impact the environment and take appropriate steps to prevent or mitigate damage.

ELEMENTS OF ENVIRONMENTAL PLANNING

D-11. The staff should plan the operation to achieve mission objectives while minimizing the environmental effects and observing environmental requirements. Although not all of the following elements are applicable to all operations (some, such as identification of alternatives to obtaining objectives, are not required for operations overseas), they may prove helpful during planning:

- Identifying operational objectives and the activities that are proposed to obtain these objectives, including logistics and the identification of HAZMAT that may be used.
- Identifying potential alternative means of obtaining operational objectives. Alternatives may include the use of new technologies to minimize the impact on the environment.
- Identifying the environmental requirements applicable to the operational area.
- Identifying adverse environmental health and environmental impacts that may result from the operation.
- Establishing formal relationships and coordinating with other disciplines that have roles in environmental planning and operations (medical and legal).
- Identifying the environmental characteristics of the potentially affected area.
- Identifying possible environmental contingencies that may occur during the operation, such as accidental spills.
- Determining how the environmental contingency would affect the environment in the operational area and how it could be prevented or mitigated should it occur.
- Determining the environmental and operational risk associated with the operation. If risks are unacceptable, identifying alternatives that mitigate associated risks.
- Establishing standardized identification (signage or markings) of off-limits or high-risk areas.
- Negotiating applicable agreements to allow for the unimpeded transit of HAZMAT or waste by military and contracted assets for environmentally sound treatment, storage, or disposal.
- Determining contractor status, to include privileges and immunities, in support of the operation.
- Identifying environmental resources and reachback capabilities.

KEY ENVIRONMENTAL FACTORS

D-12. Commanders should consider environmental and FHP during each phase of an operation. In planning and conducting military operations, regardless of geographic location, the engineer supports the commander in giving appropriate consideration to the following:

- Environmental conditions that preexist and impact site selection and environmental health vulnerabilities, as well as potential U.S. liabilities associated with the operation.
- Air emissions.
- HAZMAT, including pesticides.
- Hazardous waste. Appropriate disposition could include recovery, treatment, or disposal within the operational area or, where necessary, transportation to another location for these purposes.
- Oil and hazardous substance spill prevention, control, and response training.
- Medical and infectious waste.
- Solid waste.
- Soil contamination.
- Water and wastewater, to include sanitary wastewater.
- Natural resources, to include endangered or threatened species and marine mammals.
- Historic and cultural resources.
- Noise abatement.
- Resource and energy conservation through pollution prevention practices.
- Camp closure and site cleanup before redeployment.
- Incident reporting and documenting any cleanup action.
- Excess material and equipment being transported from the tactical area in an environmentally sound manner.
- Contractors and their vehicles having unhindered transit across international borders.

RISK MANAGEMENT

D-13. FM 5-0 describes risk management as the process of detecting, assessing, and controlling risk arising from operational factors and balancing risk with mission benefits. Risk management is an integral part of the MDMP. FM 5-480 outlines the multi-Service risk management process and provides the framework for making risk management a routine part of planning, preparing, and executing operational missions and everyday tasks. FM 3-100.4 further clarifies the elements of risk when focused on environmental considerations. Assessing environmental-related risks is part of the total risk management process. The knowledge of environmental factors is key to planning and decision making. With this knowledge, leaders quantify risks, detect problem areas, reduce the risk of injury or death, reduce property damage, and ensure compliance with environmental laws and regulations. Unit leaders should conduct risk assessments before conducting any training, operations, or logistical activities.

Appendix E

Base Camp Estimating and Planning Considerations

The purpose of this appendix is to provide the engineer with some basic common planning factors and standards for base camp construction. Additional information may be found in JP 3-34, the TMs that support TCMS, FM 101-10-1, and other documents as identified throughout this appendix.

DEFINITIONS

STANDARDS OF CONSTRUCTION

E-1. The planning factor for the standards of construction are based on the expected duration of the contingency and described in JP 3-34. Standards of construction as are as follows:

- Organic (0 to 90 days).
- Initial (0 to 24 months).
- Temporary (6 months to 5 years).
- Semipermanent (5 years to 10 years).

CLIMATE ZONE

E-2. There are four different climatic zones, with the temperate being the base. The work effort for other zones is determined by applying the following factors: tropical (1.45), desert (1.15), and frigid (2.57).

QUALITY OF LIFE

E-3. The quality of life captures the operational commander's requirements for bed-down and base camp living standards.

BASE CAMP CONSTRUCTION ESTIMATING

E-4. Table E-1 and Table E-2, page E-2, provide a summary of the engineer work effort, land requirements, key materials, and utilities necessary for various size base camps. The tables are designed to be used during initial planning for base camp construction.

Table E-1. Summary table, base camp engineer construction effort

Base Camp Size	Short Tons	Equipment Hours	Man-Hours			
			Horizontal	Vertical	General	Total
500	2,755	77	3,506	33,175	10,232	46,913
1,500	7,698	247	8,124	86,047	26,331	120,502
3,000	15,138	503	15,093	171,012	53,730	240,070
10,000	50,460	1,680	51,093	570,040	179,100	800,233

Table E-2. Summary table, base camp area, aggregate, and utilities requirements

Base Camp Size	Real Estate Acre	Fine Aggregate (cu yd)	Course Aggregate (cu yd)	Potable Water (GPD)	Sewage (GPD)	Electricity (kW)
500	16.0	450	620	12,500	8,750	182
1,500	51.4	1,700	2,485	37,500	26,250	486
3,000	104.7	3,320	4,820	75,000	52,500	988
10,000	350	11,200	16,066	250,000	175,000	3,293

GENERAL CONSTRUCTION EFFORT REQUIREMENTS

E-5. Tables E-3 and E-4 describe the general construction effort requirements necessary for typical site preparation and basic facilities for a 500-man base camp.

Table E-3. Construction effort, site preparation requirements

Facility Description	Size (sq yd)	Basis	Qty	Man-Hours			
				Hor	Ver	Gen	Total
Road, Class A, 1-inch multisurface, 1-mile	—	as required	0.2	58	NA	10	68
Hardstand	1,000	as required	4.0	168	NA	80	248
Road, Class A, graded and drained		as required	0.2	235	NA	84	319
Hardstand	1,000	as required	4.0	288	NA	104	392
Site preparation, 1-acre	—	as required	5.0	440	NA	160	600
TOTAL				3,506	33,175	10,232	46,913

Table E-4. Construction effort, facilities requirements (temporary to semipermanent standard, temperate climate, or wood frame)

Facility Description	Size	Basis	Quanity	Man-Hours			
				Hor	Ver	Gen	Total
Shop, motor repair	48 x 48 x 14 ft	1 per 100 vehicles	1	55	1,185	287	1,527
Storehouse	20 x 50 x 8 ft	2 sq ft per man	1	32	461	136	629
Dispensary	20 x 60 x 8 ft	1 per 500 men	1	33	1,290	115	1,438
Headquarters and unit supply	20 x 40 x 8 ft	1 per 200 men	3	84	1,293	240	1,617
Barracks, 500-man	20 x 100 x 8 ft	40 sq ft per man	10	450	7,510	1,860	9,820
Kitchen	—	1 per 250 men	2	154	10,352	3,788	14,294
Bathhouse and latrine	20 x 30 x 8 ft	1 shower per 10 men	1	24	941	61	1,026
Bathhouse and latrine	20 x 80 x 8 ft	1 shower per 24 men	1	39	1,754	150	1,943
Quarters (officer)	20 x 100 x 8 ft	80 sq ft per officer	1	45	869	186	1,100
Guard house	20 x 60 x 8 ft	1 to 250 men	1	33	626	115	774
Day room	40 x 60 x 8 ft	5 sq ft per man	1	43	868	178	1,089
Electric distribution	500-man	light and power	1	56	460	192	708
Boiler plant	—	1/2 mess	1	208	4,112	1,200	5,520
Drainage	500-man	17.5 GPD	1	205	384	490	1,079
Water supply well	—	as required	1	396	45	230	671
Water tank	200 gallons	as required	1		105	4	109
Water distribution	500-man	25 GPD per man	1	352	812	416	1,580
Sump fire	10,000 gallons	effective radius 500 feet	1	16	108	16	240

SUPPORT AND STORAGE FACILITIES CONSTRUCTION PLANNING FACTORS

E-6. Tables E-5 and E-6, page E-4, provide construction planning information on motor parks and troop support facilities.

Table E-5. Motor park

Base Camp Size	Area (sq ft)
500	61,760
1,500*	242,160
3,000*	541,200
10,000*	721,600
*Based on combinations of 500-, 1,000-, and 5,000-man estimates.	

Table E-6. Troop support facilities

Description	Units	Criteria	500	1,500	3,000	10,000
Dining facility	sq ft	sq ft per person varies by unit size	14.0	11.0	11.0	11.0
Fire station	sq ft	2.6 x size of vehicle + 90 sq ft	—	—	—	—
I/R facility	sq ft	250 sq ft military police + 50 sq ft per confinee	—	—	—	—
Bakery	sq ft	0.6 sq ft per person supported	300.0	900.0	1,800.00	6,000.0
Laundry	sq ft	sq ft per person varies by unit size	4.4	4.4	3.30	3.0
Dry cleaning	sq ft	sq ft per person varies by unit size	4.4	4.4	1.75	1.0
Chapel	sq ft	1.785 sq ft per person	893.0	2,678.0	5.55	17,850.0
Craft and hobby	sq ft	1.0 sq ft per person	500.0	1,500.0.0	3,000.00	10,000.0
Gymnasium	sq ft	3.3 sq ft per person	1,650.0	4,950.0	9,900.00	33,000.0
Library	sq ft	0.75 sq ft per person	375.0	1,125.0	2,250.00	7,500.0
Service club	sq ft	7.5 sq ft per NCO; 9.5 sq ft per officer	—	—	—	—
PX	sq ft	1.2 sq ft per person	600.0	1,800.0	3,600.00	12,000.0
Post Office	sq ft	sq ft per person varies by unit size	NA	NA	0.50	0.5
Theater	sq ft	sq ft per person varies by unit size	NA	NA	5.50	5.5

E-7. Examples of selected storage requirements and planning factors for base camps are addressed in tables E-7, E-8, and E-9.

Table E-7. Covered and open storage requirements for 14 days of stockage

Base Camp Size	Covered Storage (sq ft)	Open Storage (sq yd)
500	44	1,330
1,500	132	3,990
3,000	265	7,980
10,000	882	26,600

Table E-8. Cold storage requirements for 14 days of stockage

Base Camp Size	Class I (cu ft)	Class VI (cu ft)	Class VIII (cu ft)	Class IX (cu ft)
500	585	155	34	12
1,500	1,755	465	101	36
3,000	3,510	930	202	72
10,000	11,690	3,095	672	238

Table E-9. Fuel storage

Base Camp Size	Diesel Barrels (bbl)*	MOGAS (bbl)*
500	160	600
1,500	480	1,800
3,000	960	3,600
10,000	3,200	12,000
*Assuming stock objective of 8 days.		

TROOP HOUSING

E-8.   Tables E-10, E-11, and E-12 provide basic planning guidance for troop housing.

Table E-10. Troop housing

Base Camp Size	Officer (sq ft)	Enlisted (sq ft)
500	11,000	28,800
1,500	33,000	86,400
3,000	66,000	172,800
10,000	220,000	576,000
Note. Assumes 20/80 officer to enlisted ratio; 110 sq ft per officer; 72 sq ft per enlisted		

Table E-11. Quality-of-life standards for tentage

Tier Level	Bed-Down and Base Camp Living Standards
Tier I	Simple tent setup without floor, nonpermanent
Tier II	Wooden floor, lights, pole-supported, 2 electrical outlets
Tier III	Slightly nicer wooded floor, 2/3 wooden wall structure with frame, more electrical outlets

Table E-12. Selected tentage planning factors

Type	Floor Area (sq ft)	Weight Packed (lb)	Volume Packed (cu ft)
Tent, GP, small	198.9	163	26.2
Tent, GP, medium	512.0	534	33.0
Tent, GP, large	936.0	665	69.0
Tent, ext modular (temper)	640.0	2,192	200.0
Tent, maintenance, medium	640.0	1,798	62.0
Note. Operation Joint Endeavor living standard was 10 soldiers per GP medium.			

ADDITIONAL PLANNING CONSIDERATIONS

E-9. General planning factors for water, utilities, and transportation requirements are listed in tables E-13, E-14, and E-15.

Table E-13. General planning factors for potable and nonpotable water requirements

Consumer	Rate of Consumption	Remarks
Individual	3 to 6 GPD per man	—
Camp (initial with bath)	25 to 50 GPD per man	Include waterborne sewage
Vehicles (tactical)	1/2 to 1 GPD per vehicle	—
Support Facilities		
Hospital	200 GPD per bed	20-hour operation
QM laundry company	64,000 GPD	20-hour operation
Construction Equipment		
Road construction	10,000 G/km	Nonpotable, clean
Rock crusher	22,500 GPH	Nonpotable, clean
Concrete mixer	560/140 GPH	Nonpotable, clean
Other Considerations		
Sewage treatment requirements	2.5 gallons per man per day	Nonpotable, clean
Garbage (food waste)	2.5 gallons per man per day	Nonpotable, clean
Refuse (other waste)	—	Nonpotable, clean

Table E-14. General planning factors for electrical power and distribution requirements

Facility	Electrical Power and Distribution Requirements
Installation	0.7 kW per man
Hospital	1.6 kW per bed

Table E-15. Selected transportation information

| Air[1] | | | Sea | | Land | | | |
Aircraft	Allowable Cabin Load (lb)	Allowable (cu ft)	Ship or Barge	Capacity (long ton)	Motor	Load (short ton per trip)	Rail[2]	Usable Cube (short ton)
C-5A	204,000	18,368	7,029	2,207.8	2.5-ton	2.5	Well flatcar	50
C-141	90,200	7,024	7,028	1,131.2	5-ton	5.0	Medium flatcar	25
C-130	35,000	2,818	7,005	570	12-ton S&P	12.0	Small flatcar	12
C-17	167,000		231 A	585	22.5-ton flatbed	15.0	Boxcar	10
KC-10A	169,350	12,980	231 B	578	34-ton trailer	25.0	Coaches	40 troops
B-747	180,000		2,001	24-hour operation for troops/ 24 passenger	60-ton semitrailer	40.0	Sleepers	32 troops

[1]Estimates are for peacetime payload planning.
[2]Maximum length of a train is 40 cars; maximum net load is 400 tons or 1,000 troops.

PRELIMINARY ESTIMATES

E-10. The preliminary estimate performed in planning for base camp construction should contain the following:

- Real estate required, in area (square feet or meters).
- Equipment hours required for construction.
- Man-hours required for construction, by construction skill.
- Materiel requirements in short tons.

BASE CAMP MASTER PLANNING

E-11. Master planning is a comprehensive process that brings together a variety of personnel and their requirements. It addresses the need to comprehensively plan future facilities to meet various and competing interests in support of the mission. The administrative control relationships and C2 may vary from one contingency operation to the next. The discussion in this appendix is based on the assumption that a task force is in place and is responsible to a combatant command for executing all Title 10 responsibilities in its AO.

- The combatant commands role in master planning is to provide standards and appropriate factors oversight of the base camp development process and approve the expenditure of funds according to established policy (for example, joint acquisition review board instructions).
- Task force commanders establish base camp planning boards at each base camp and forward operating site. They also provide guidance on conducting planning boards, issue commander guidance, and coordinate master planning components.

E-12. A base camp master plan is the commander's comprehensive plan for the orderly and efficient management and development of land, facilities, and infrastructure to support the mission.

E-13. Most base camps and forward operating sites will establish and run a base camp planning board and develop a base camp master plan. Some smaller or remote forward-operating sites may fall under the command of a larger base camp or forward-operating site that provides this support and oversight.

E-14. The development and use of a base camp master plan can enhance operational area security, operational readiness, personnel safety, the efficient use of resources, living conditions, and the quality of life experienced during the contingency operations. Proper zoning and improvements to the facilities and utilities and the efficient investment of resources will increase the quality of life for all Soldiers while enhancing their physical protection.

E-15. An organized site plan is a crucial part of a master plan. A well-designed site layout will address at least the following:

- Biological factors that may affect physical health.
- Financial factors that could reduce operational and renovation costs.
- Psychological factors that affect attitudes.
- Social factors, including coordination and cooperation.

E-16. The base camp planning board provides the forum for base camp and forward-operating site managers and leaders to make comprehensive, balanced decisions on the future layout of forward-operating site facilities and infrastructure. The board should meet periodically to review and refine plans based on changing mission priorities, with the goal of providing high-quality living and working environments appropriate to the current mission intent while incorporating best business practices.

MASTER PLAN COMPONENTS

E-17. Typical master plans consist of long- and short-range components and a capital investment strategy.

LONG-RANGE COMPONENTS

E-18. The long-range component is an assessment of what the base camp or forward-operating site should look like in an extended period of time (for example, 5 years). The process usually includes infrastructure, transportation flow, zoning, aesthetics, and signage requirements. The following items that could make up the long-range components of a master plan include—

- Long-range analysis.
- Environmental-baseline analysis.
- Utilities assessment.
- Transportation assessment.
- Land-use analysis and zoning plan.
- Physical security plan (overlay).
- Fire protection plan (overlay).
- Base camp and forward-operating site design guidelines.
- Capacity expansion analysis.
- Supporting graphics and overlays.
- Ammunition holding area explosive quantity distance site plan.

SHORT-RANGE COMPONENTS

E-19. The short-range component of a master plan is the immediate or temporary solution to facility imbalances to be used until more permanent solutions are found. Temporary solutions may include relocations or temporary diversions in uses of facilities and temporary construction. The short-range component includes site-specific graphics with locations of projects. The following items could make up the short-range component:

- An assessment of how to "get well," analyses of alternatives, evaluation and selection of preferred alternatives, and narrative justification for the selected COA.
- Assets and facilities investment plan.

- Environmental documentation.
- Assets disposal list.
- Supporting graphics.

E-20. The base camp planning board acts as the commander's "board of directors" to ensure the orderly development and management of base camp or forward-operating site facilities (and supported sites). It guides the development and maintenance of all components of the base camp and forward-operating site. It should coordinate base camp and forward-operating site planning with the following:

- An adjacent or nearby site.
- Affected host-nation agencies.

E-21. The BCPB ensures that the base camp master plan does the following:

- Addresses facility requirements for all activities of the base camp or forward-operating site and supported areas.
- Projects for growth or reduction in units and activities assigned to the base camp or forward operating site based on changes in the mission.
- Determines site-specific base camp or forward-operating site design guidelines, and adheres to standards.
- Reviews funding projections, advises the commander of priorities, and proposes COAs.
- Ensures maximum efficient use of existing facilities.
- Ensures that project plans and projects are consistent with good environmental stewardship.
- Makes recommendations on space utilization.

CAPITAL INVESTMENT STRATEGY

E-22. The capital investment strategy analyzes shortfalls and excesses in facilities through a tabulation of existing and required facilities and identifies preferred action plans to solve imbalances. The recommendations (solutions) must be consistent with the long-range component.

FACILITIES STANDARDS

E-23. The CCDR will specify the construction standards for facilities in the theater (see paragraph 11-9). As mentioned previously, the intended life span of the facilities and infrastructure of a base camp or forward-operating site will depend on mission-driven and economic decisions. There are three sets of construction standards used that are determined by the expected base camp and forward-operating site life span. During the life cycle of a base camp or forward-operating site, authorized facilities may progress from initial to semipermanent or may be immediately established at any level depending on operational requirements. Meeting established facility standards may be a progressive effort; however, CCDRs will strive to meet their standards as quickly as the operational situation permits. Table E-16, pages E-10 through E-15, provides an example of possible standards for initial, temporary, and semipermanent facilities for base camps and forward-operating sites. Permanent standards of construction have a life expectancy of greater than 10 years. The CCDR must specifically approve permanent construction.

Table E-16. Example of initial, temporary, and semipermanent facility standards

Facility	Initial (Less Than 6 Months)	Temporary (6 Months to Less Than 24 Months)	Semipermanent (2 Years to Less Than 10 Years)
American forces network-manned operations	None	Container, SEAhut	Container; SEAhut; metal, prefabricated building
American forces network-unmanned operations	None	Container, SEAhut	Container; SEAhut; metal, prefabricated building
Alteration/pressing shop	None	Tier III tents, SEAhuts, containers	SEAhuts, containers: 2 to 10 years Masonry and prefabricated buildings: 10 or more years
ASG, area support team	None	Tier III tents, SEAhuts, containers	SEAhuts and containers: 2 to 10 years Masonry and prefabricated buildings: 10 or more years
ASP	Containers	Containers to bunkers	Bunkers
Athletic fields	None	Grassed fields	Grassed fields with lights
Aviation fuel	HEMTT tanker	Bladder	Metal tanks, steel lines
Aviation maintenance	Organic tentage, force provider[1]	Aviation clamshell tent with sand-filled plywood, asphalt, or concrete floor	Aviation clamshell tent with sand-filled plywood, asphalt, or concrete floor
Barber shop, beauty shop	None	Tier III tent, SEAhuts, containers	SEAhuts, containers: 2 to 10 years Masonry and prefabricated buildings: 10 or more years
Basic load ammunition holding area, captured ammunition holding area	Military vans (container) with earth berms	Earth-covered, standard, steel-reinforced bunkers on concrete pads with berms	Earth-covered, standard, steel-reinforced bunkers on concrete pads with berms
Chapel	Organic tentage with wooden floors, Tier I tents, "Chapel-in-a-Box", force provider[1]	SEAhut, containers	Davidson-like, wood-frame building; SEAhuts; containers: 2 to 10 years Masonry and prefabricated buildings: 10

Table E-16. Example of initial, temporary, and semipermanent facility standards

Facility	Initial (Less Than 6 Months)	Temporary (6 Months to Less Than 24 Months)	Semipermanent (2 Years to Less Than 10 Years)
			or more years
Cold storage	Portable refrigeration with freezer units for medical, food, and maintenance storage	Refrigeration installed in temporary structures	Refrigeration installed in semipermanent structures: may be preengineered buildings
Communications compound, national service center	Organic tentage with wooden floors, Tier I tents, force provider[2]	Tier III tents, SEAhuts, containers	SEAhuts and containers: 2 to 10 years Masonry and prefabricated buildings: 10 or more years
Community activity center	None	SEAhuts	SEAhuts: 2 to 10 years Masonry and prefabricated buildings: 10 or more years
Dining facility	Mobile kitchen trailer, organic tentage with wooden floors, Tier I tents, personnel protection[1]	Tier III tents, SEAhuts, fest tents	SEAhuts: 2 to 10 years Masonry and prefabricated building: 10 or more years
Defense Reutilization and Marketing Office	None	Metal, prefabricated building with concrete or asphalt floor and gravel holding area	Metal, prefabricated building with concrete or asphalt floor with gravel holding area
DS maintenance	Organic tentage or force provider[2]	Metal, two-story prefabricated building on concrete base with concrete aprons	Metal, two-story prefabricated building on concrete base with concrete aprons
Direct exchange, central issue facility	None	Tier III tents, SEAhuts, containers, metal prefabricated building	SEAhuts and containers: 2 to 10 years Masonry and prefabricated buildings: 10 or more years
Education center	None	Tier III tents, SEAhuts, containers, metal prefabricated building	SEAhuts, containers: 2 to 10 years Masonry and prefabricated buildings: 10 or more years

Table E-16. Example of initial, temporary, and semipermanent facility standards

Facility	Initial (Less Than 6 Months)	Temporary (6 Months to Less Than 24 Months)	Semipermanent (2 Years to Less Than 10 Years)
Electrical	Tactical generators with high- and low-voltage distribution, organic equipment, force provider[1]	Commercial power with nontactical power and high- or low-voltage distribution backup	Commercial power with nontactical power and high- or low-voltage distribution backup
Field house, multipurpose facility	None	Metal, prefabricated building	Metal prefabricated building
Finance and personnel support operations	None	Tier III tents, SEAhuts, containers	SEAhuts and containers: 2 to 10 years Masonry and prefabricated buildings: 10 or more years
Fire protection	Organic equipment, portable fire extinguishers	See paragraph 11-63.	See paragraph 11-63.
Fitness center	None	SEAhuts; metal, prefabricated building	SEAhuts: 2 to 10 years Masonry and prefabricated buildings: 10 or more years
Ground fuel	Organic equipment, bags, force provider with secondary containment	Bladders with secondary containment	Metal tanks with steel lines with secondary containment
HAZMAT warehouse	Storage container	SEAhuts or metal, prefabricated building with secondary containment	SEAhuts and metal, prefabricated buildings and secondary containment: 2 to 10 years Masonry and metal, prefabricated buildings with secondary containment: 10 or more years
Hazardous waste	Storage container, removal from theater	Covered, built-on elevated pad with secondary containment (civilian contract removal)	Covered, built-on elevated pad with secondary containment (civilian contract removal)
Helipad	Tactical surfacing, including matting	Concrete with aprons	Concrete with aprons

Table E-16. Example of initial, temporary, and semipermanent facility standards

Facility	Initial (Less Than 6 Months)	Temporary (6 Months to Less Than 24 Months)	Semipermanent (2 Years to Less Than 10 Years)
Housing	Organic tentage with wooden floors, Tier I tents, force provider[1]	Tier III tents, SEAhuts, containers	SEAhuts and containers: 2 to 10 years Masonry and prefabricated buildings: 10 or more years
Kennel	Organic tentage, Tier I tents (DA Pamphlet 190-12)	SEAhuts, container-adapted to DA Pamphlet 190-12 criteria	SEAhuts and containers adapted to DA Pamphlet 190-12 criteria
Latrines and septic systems	Organic equipment, evaporative ponds, pit burnout latrines, lagoons for hospitals, force provider[1]	Waterborne from ablution units or SEAhuts to austere treatment facility	Waterborne to wastewater treatment plant from SEAhuts and ablution units: 2 to 10 years Masonry and prefabricated buildings: 10 or more years
Laundry collection and distribution point	Organic tentage with wooden floors, Tier I tents, force provider[1]	Tier III tents, SEAhuts, containers	SEAhuts and containers: 2 to 10 years Masonry and prefabricated buildings: 10 or more years
Medical (See paragraph 11-24 for further guidance.)	Organic tentage with wooden floors, medical tents, Tier I tents	SEAhuts; medical metal, prefabricated buildings; refrigerated containers	SEAhuts; medical metal, prefabricated buildings: 2 to 10 years Masonry and medical, metal prefabricated buildings: 10 or more years
Medical waste	Field incinerator	Incinerator, civilian contract	Incinerator, civilian contract
Military police station	Organic tentage with wooden floors, Tier I tents, force provider[1]	Tier III tents, SEAhuts, containers	SEAhuts and containers: 2 to 10 years Masonry and prefabricated buildings: 10 or more years
Morgue	Refrigerated container	SEAhut, container with Gortex for private fencing, refrigerated container	SEAhuts and containers: 2 to 10 years Masonry and prefabricated buildings: 10 or more years
Multipurpose theater	None	Metal, prefabricated building	Metal, prefabricated buildings

Table E-16. Example of initial, temporary, and semipermanent facility standards

Facility	Initial (Less Than 6 Months)	Temporary (6 Months to Less Than 24 Months)	Semipermanent (2 Years to Less Than 10 Years)
MWR warehouse, maintenance facility	None	Metal, prefabricated building	Metal, prefabricated buildings
Nonpotable water	Local source	Local source	Local source
Office	Organic tentage with wooden floors, Tier I tents, FP[1]	Tier III tents, SEAhuts, containers	SEAhuts and containers: 2 to 10 years Masonry or prefabricated buildings: 10 or more years
Parking lots	Gravel	Gravel with concrete turning pads for tracked vehicles	Gravel with concrete turning pads for tracked vehicles
Perimeter fence	Triple standard	USACE Standard FE6	USACE Standard FE6
Perimeter lights	Generator sets	Fixed lighting	Fixed lighting
Postal	None	Metal, prefabricated building	Metal, prefabricated building
PX	AAFES trailer	Davidson-like, wood-frame building; metal prefabricated building	Metal, prefabricated building
Post warehouse	AAFES trailer	Davidson-like, wood-frame building; container; metal prefabricated building	Metal, prefabricated building
Potable water	Bottled water or water points, wells, other potable-water production and pressurized water distribution systems, reverse osmosis water purification unit, force provider	Wells, treatment plants	Wells, treatment plants
Road	Gravel	Gravel	Primary roads: asphalt with concrete turning pads Secondary and perimeter patrol roads: gravel
Runway and taxiway	Tactical surfacing, including aggregate and stabilized earth	Paved	Paved

Table E-16. Example of initial, temporary, and semipermanent facility standards

Facility	Initial (Less Than 6 Months)	Temporary (6 Months to Less Than 24 Months)	Semipermanent (2 Years to Less Than 10 Years)
Shower	Organic equipment, personnel protection[1]	Ablution units or SEAhuts	SEAhuts and AB units: 2 to 10 years Masonry or prefabricated buildings: 10 or more years
Solid waste	Field incinerator	Incinerator, civilian contract and recycling when possible	Incinerator, civilian contract, recycling program, composting
Squadron operations building	Organic tentage with wooden floors, Tier I tents, force provider[1]	SEAhuts, metal prefabricated building	SEAhuts and metal prefabricated buildings: 2 to 10 years Masonry and metal prefabricated buildings: 10 or more years
Supply support activity warehouse	Organic tentage with wooden floors, Tier I tents, force provider[1]	Metal, prefabricated building	Metal, prefabricated building
Training facilities	None	See paragraph 11-64.	See paragraph 11-64.
Vehicle maintenance	Organic tentage, force provider[1]	Metal, two-story, prefabricated building on concrete base with concrete aprons	Metal, two-story, prefabricated building on concrete base with concrete aprons
Washrack	Gravel lot	Gravel lot with oil-water separator and gray-water discharge	Elevated, flat, and container rack with oil-water separator and gray-water discharge

[1]Force provider: Each force provider module supports 550 personnel, plus 50 operators with climate-controlled billeting (with planning factors of 15 Soldiers per tent); food service (1,800 A-rations meals per day); laundry service (200 pounds per hour); showers and latrines (one 10-minute shower per day); MWR facilities and equipment; power (60-kilowatt tactical quiet generators [1.1. megawatts continuous]); prime power connection kit; water storage and distribution (80,000 gallons for every 3 days); fuel storage and distribution (20,000 gallons for every 3 days); waste-water collection (30,000 gallons per day); and system support packages (30 days spare and repair parts).

This page intentionally left blank.

Source Notes

This section lists sources by page number. Where material appears in a paragraph, both the page number and paragraph number are listed. Boldface indicates titles of vignettes.

1-1   "Although they were the size of David, engineers did the work of Goliath." Assistant Division Commander, 101st Airborne Division (Air Assault), Operation Iraqi Freedom After-Action Review.

2-1   "Above all, we must realize that no arsenal, or no weapon in the arsenals of the wicked are so formidable as the will and moral courage of free men and women. It is a weapon adversaries in today's world do not have." Ronald Reagan [Online]. Available: <http://www.quotationspage.com>.

3-1   "Never tell people how to do things. Tell them what to do and they will surprise you with their ingenuity." General George Patton [Online]. Available: <http://www.brainyquote.com>.

4-1   "Seek first to understand, then to be understood." Stephen R. Covey [Online]. Available: <http://www.learnoutloud.com/catalog/business/strategy>.

5-1   "They must float up and down with the tide. The anchor problem must be mastered. Let me have the best solution worked out. Don't argue the matter. The difficulties will argue for themselves." Winston Churchill on pier construction to support the invasion, May 1943 [Online]. Available: <http://www.britannica.com/dday/article-9344572>.

6-1   "Air power is a thunderbolt launched from an eggshell invisibly tethered to a base." Hoffman Nickerson, Arms and Policy (1945) [Online]. Available: <http://www.quote-fox.com>.

7-1   "The line that connects an army with its base of supplies is the heel of Achilles—its most vital and vulnerable point." John S. Mosby, Colonel, Confederate States of America (1887) [Online], Available: <http://grapevinedispatches.wordpress.com/category/john-s-mosby/>.

8-1   "History shows that army campaigns in undeveloped countries have often involved waging war against natural obstacles, rather than against a foe." Air Marshall E. J. Kingston-McCloughry.

9-1   "The art of war teaches us to rely not on the likelihood of the enemy not coming, but on our readiness to receive him not on the chance of his not attacking, but rather on the fact that we have made our position unassailable." Sun Tzu [Online]. Available: <http://chinapage.com>.

10-1   "Class IV stocks should be robust and ready for crisis projects. If engineers don't stock Class IV, no one else will." S-4, 130th Engineer Brigade, Operation Iraqi Freedom After-Action Review.

11-1   If you can figure out the criteria for base camp selection…you've done something the Army can use." LTG Robert B. Flowers, 50th Chief of Engineers [Online]. Available: http://www.usma.edu/publicaffairs/PV/011129/Systems.htm

12-1   "Simplicity is the ultimate sophistication." Leonardo da Vinci [Online]. Available: <http://www.brainyquote.com>.

13-1   "Except when war is waged in the desert, noncombatants, also known as civilians or "the people," constitute the great majority of those affected." Martin Van Creveld [Online]. Available: <http://www. Brainy.quote.com>.

14-1   "One should know one's enemies, their alliances, their resources and nature of their country, in order to plan a campaign."Frederick the Great[Online]: <www.airpower.au.af.mil/airchronicles/aureview/1977/mar-apr/papworth.html>.

15-1   "...before the shooting begins. The bravest men can do nothing without guns, the guns nothing without ammunition; and neither guns nor ammunition are of much use in mobile warfare unless there are vehicles with sufficient petrol to haul them around." Field Marshal Erwin Rommel, World War II [Online]: <www.quartermaster.army.mil/OQMG/Professional_Bulletin/1998/Spring_1998/ipds.html>.

16-1   "I don't know what the hell this 'logistics' is that Marshall is always talking about, but I want some of it." Fleet Admiral Ernest J. King, 1942 [Online]. Available: <http://www.encyclopedia.com>.

Glossary

The glossary lists acronyms/abbreviations and terms with Army or joint definitions, and other selected terms. Where Army and joint definitions are different, (Army) follows the term.

SECTION I – ACRONYMS AND ABBREVIATIONS

AA	assembly area
AAFES	Army and Air Force Exchange Service
AC	alternating current
ACS	Assistant Chief of Staff
ADC	area damage control
admin	administrative
ADP	automatic data processing
ADR	airfield damage repair
AFCESA	Air Force Civil Engineering Support Agency
AFCS	Army Facilities Components System
AFJPAM	Air Force Joint Pamphlet
AFM	Air Force manual
AFPAM	Air Force pamphlet
AFTTP	Air Force Technical Training Publication
AMPHIB	amphibian
AO	area of operations
AOR	area of responsibility
APOD	aerial port of debarkation
APOE	aerial port of embarkation
AR	Army regulation
ARNG	Army National Guard
ARNGUS	Army National Guard of the United States
ARRK	automated route reconnaissance kit
ART	Army tactical task
ASCC	Army service component commander
ASG	area support group
ASP	ammunition supply point
AT	antiterrorism
ATCALS	air traffic control and landing system
ATP	ammunition transfer point
ATTN	attention
AUTL	Army Universal Task List

AutoCAD®	automated computer-assisted design
AVLB	armored vehicle-launched bridge
bbl	barrels
BCT	brigade combat team
BDT	base development team
BG	bag
bn	battalion
BOM	bill of materials
BPA	blanket purchase agreement
BT	bomb trench
C2	command and control
C3	command, control, and communications
CA	civil affairs
CADD	computer-aided design and drafting
CALL	Center for Army Lessons Learned
CBMU	construction battalion maintenance unit
CBR	California bearing ratios
CBRN	chemical, biological, radiological, and nuclear
CBRNE	chemical, biological, radiological, nuclear, and high-yield explosives
CBT	combat
CCD	camouflage, concealment, and deception
CCDR	combatant commander
CCP	casualty collection point
CEWL	corps engineer work line
CFM	cubic feet per minute
CI	civilian internee
CJCSM	Chairman of the Joint Chiefs of Staff manual
CL	class
CM	consequence management
CMMC	corps materiel management center
CMOC	civil-military operations center
CMU	concrete masonry unit
co	company
COA	course of action
COCOM	combatant command (command authority)
COL	colonel
COMMZ	communications zone
CONEX	container express
CONPLAN	concept plan
CONUS	continental United States
COP	common operational picture

COS	Chief of Staff
COTS	commercial off-the-shelf
CREST	contingency real estate support team
CSH	combat support hospitals
CU	cubic
CUCV	commercial utility, cargo vehicle
cu ft	cubic foot (feet)
cu yd	cubic yard(s)
CW3	Chief Warrant Officer, W-3
CW4	Chief Warrant Officer, W-4
CW5	Chief Warrant Officer, W-5
DA	Department of the Army
DC	dislocated civilian
DCP	dynamic cone penetrometer
DD	Department of Defense
DENIX	Defense Environmental Network and Information Exchange
DEPMEDS	Deployable Medical Systems
DEPORD	deployment order
DHS	Department of Homeland Security
dist	district
div	division
DMMC	Division Materiel Management Center
doc	document
DOD	Department of Defense
DOS	Department of State
DOTMLPF	doctrine, organization, training, materiel, leadership and education, personnel, and facilities
DPGDS	deployable power generation and distribution system
DODI	Department of Defense Instruction
DS	direct support
DSB	dry support bridge
DX	directed center of expertise
E1	Private 1
E2	Private 2
E3	Private First Class
E4	Specialist
E5	Sergeant
E6	Staff Sergeant
E7	Sergeant First Class
E8	Master Sergeant; First Sergeant
E9	Sergeant Major; Command Sergeant Major

ea	each
EBS	environmental baseline survey
ECF	entry control facility
ECP	entry control point
ECU	environmental control unit
EH	explosive hazards
EHSA	environmental health site assessment
EI2RC	Engineering Infrastructure and Intelligence Reachback Center
ENCOM	engineer command
ENCOORD	engineer coordinator
engr	engineer
EO	executive order
EOD	explosive ordnance disposal
EPW	enemy prisoner of war
ERDC	Engineering Research and Development Center/Laboratories
ESB	engineer support battalion
ETL	engineering technical letter
EWL	engineer work line
F	Fahrenheit
FACEDAP	Facility and Component Explosive Damage Assessment Program
FARP	forward arming and refueling point
FEMA	Federal Emergency Management Agency
F&ES	fire and emergency service
FEST	forward engineer support team
FEST-A	forward engineer support team—advanced
FEST-M	forward engineer support team—main
FFE	field force engineering
FHP	force health protection
FLD	field
FM	field manual
FOB	forward operating base
FOD	foreign object damage
FRAGO	fragmentary orders
FSCOORD	fire support coordinator
FSSP	fuel system supply point
ft	feet; foot
FY	fiscal year
G-2	Assistant Chief of Staff, Intelligence
G-3	Assistant Chief of Staff, Operations and Plans
G-4	Assistant Chief of Staff, Logistics
G/km	gallons per kilometer

GATER	geospatial assessment tool for engineering reachback
GB	guard bunker
GE	general engineering
gen	general
GIS	geographic information system
GP	general purpose
GPD	gallons per day
GPH	gallons per hour
GPM	gallons per minute
GPS	global positioning system
GS	general support
GSR	general support-reinforcing
HACC	humanitarian assistance coordination center
HAZMAT	hazardous material
HBCT	heavy brigade combat team
HCA	humanitarian and civic assistance
HD	hundred
HEMTT	heavy expanded mobility tactical truck
HN	host nation
HNS	host nation support
hor	horizontal
HTML	Hypertext markup language
http	Hypertext transfer protocal
HVAC	heating, ventilation, and air conditioning
HQ	headquarters
hz	hertz
IAT	infrastructure assessment team
ID	Infantry division
I/R	internment/resettlement
IED	improvised explosive devices
ike	it knows everything
IO	information operations
IPB	intelligence preparation of the battlefield
IPDS	inland petroleum distribution system (Army)
IRB	improved ribbon bridge
ISB	intermediate staging base
ISR	intelligence, surveillance, and reconnaissance
J-2	intelligence staff section
J-3	operations staff section
J-4	logistics staff section
JAB	joint assault bridge

JCS	Joint Chiefs of Staff
JEMB	Joint Environmental Management Board
JFC	joint force commander
JFLCC	joint force land component command
JLOTS	joint logistics over-the-shore
JOA	joint operations area
JP4	jet petroleum 4
JP	joint publication
JPO	Joint Petroleum Office
JTF	joint task force
Jul	July
kg	kilogram
kV	kilovolt
kVA	kilovolt amperes
kW	kilowatt
L	latrine
lb	pound
LACV-30	light air cushion vehicle-30
LARC-LX	Lighter, amphibious resupply, cargo, 60 ton
LC	landing craft
LCC	land component commander
LCM	landing craft, mechanized
LCM-8	landing craft-mechanized
LCU	landing craft utility
LED	light emitting diode
LFA	lead federal agency
LN	local national
LNO	liaison officer
LOC	line of communications
LOTS	logistics over-the-shore
LRP	load and roll pallet
LSB	logistics support brigade
LTG	lietuenant general
LZ	landing zone
m	meter(s)
M/CM/S	mobility, countermobility, and survivability
MAGTF	Marine air-ground task force
maint	maintenance
MANSCEN	Maneuver Support Center
MAOS	minimum airfield operating surface
MCRP	Marine corps reference publication

MCT	movement control team
MCWP	Marine Corps warfighting publication
MDMP	military decision-making process
MED	medical
MEF	Marine expeditionary force
MEG	means estimating guide
METL	mission-essential task list
METT-TC	mission, enemy, terrain and weather, troops and support available, time available, and civil considerations
MGB	medium girder bridge
MHE	materials handling equipment
MI	military intelligence
mil	military
MLC	military load classification
mm	millimeter(s)
mngt	management
MOB	main operating base
MOG	maximum (aircraft) on ground
MOGAS	motor gasoline
MOS	military occupational specialty
MP	military police
MSC	Military Sealift Command
MSR	main supply route
MSS	medium shelter system
MTF	medical treatment facility
MTOE	modified table of organization and equipment
MUSE	mobile utility support equipment
MW	megawatt(s)
MWR	morale, welfare, and recreation
NATO	North Atlantic Treaty Organization
NAVAIDS	navigation aid
NAVFAC	Naval Facilities Engineering Command
NAVFACINST	Naval Facility Command Instruction
NCO	noncommissioned officer
NEC®	National Electrical Code®
NESC®	National Electrical Safety Code®
NGO	nongovernmental organization
NIBS	National Institute of Building Sciences
NICP	national inventory control point
NIPRNET	Non-Secure Internet Protocol Router Network
NMCB	naval mobile construction battalion

NOTAM	notice to airmen
NSN	National Stock Number
NTTP	Navy tactics, techniques, and procedures
NWP	naval warfare publication
O1	2d Lieutenant
O2	1st Lieutenant
O3	Captain
O4	Major
O5	Lieutenant Colonel
O6	Colonel
O7	Brigadier General
O&M	operations and maintenance
OCONUS	outside the continental United States
OE	operational environment
OMEE	operation and maintenance engineering enhancement
OP	operation
OPCON	operational control
OPDS	offshore petroleum distribution system (Navy)
OPLAN	operation plan
OPNAVINST	Chief of Naval Operations Instruction
OPORD	operation order
org	organization
PM	project management
POD	port of debarkation
POL	petroleum, oils, and lubricants
PPTO	petroleum pipeline terminal operations
Prime BEEF	Prime Base Engineer Emergency Force
PSYOP	psychological operations
pt	pint
PVC	polyvinyl chloride
PX	post exchange
QM	quartermaster
QOL	quality of life
QSTAG	Quadripartite Stanardization Agreement
qty	quantity
R	reinforcing
RCT	regimental combat team
REBS	rapidly emplaced bridge system
RED HORSE	Rapid Engineers Deployable Heavy Operations Repair Squadron, Engineer
refr	refrigeration

RFF	request for forces
RFI	request for information
RO	roll
ROE	rules of engagement
RO/RO	roll-on/roll-off
ROWPU	reverse osmosis water purification unit
RP	retained personnel
RPMA	real property maintenance activities
rqn	requsition
rqmt	requirement
RSO&I	reception, staging, onward movement, and integration
RT	rough terrain
RTCH	rough terrain container handler
S-2	intelligence staff officer
S-3	operations staff officer
S-4	logistics staff
S&P	stake and platform
S&T	supply and transportation battalion
SBCT	Stryker brigade combat team
Seabee	Naval construction engineer
SEAhut	Southeast Asia hut
sec	section
SIPRNET	SECRET Internet Protocol Router Network
SITREP	situation report
SJA	Staff Judge Advocate
SME	subject matter expert
SOFA	status-of-forces agreement
SOP	standing operating procedure
SPOD	seaport of debarkation
SPOE	seaport of embarkation
sq ft	square feet
sq yd	square yard
SRB	standard ribbon bridge
SRT	special-reaction team
ST	short ton
STANAG	Standardization Agreement (NATO)
sup	supply
SVC	service
SWB	sanitary wall board
SWEAT-MSO	sewerage, water, electricity, academics, trash, medical, safety, and other considerations

TA	theater Army
TAACOM	theater Army area command
TACOM	tactical control
TACGENS	tactical generators
TAMMC	theater Army material management command
TB	technical bulletin
TC	Transportation Corps (Army)
TCE-D	tele-engineering communications equipment—deployable
TCM	theater construction manager
TCMS	Theater Construction Management System
TDA	tables of distribution and allowances
techinfo	technical information
temp	temporary
TEOC	Tele-engineering Operations Center
TETK	tele-engineering toolkit
TEWL	theater engineer work line
TG	Training guide
TM	technical manual
TO	theater of operation
TOE	table of organization and equipment
TPFDD	time-phased force and deployment data
TPFDL	time-phased force and deployment list
TPH	tons per hour
TPT	tactical petroleum terminal
TQG	tactical quiet generator
TRADOC	United States Army Training and Doctrine Command
trans	transportation
TRANSCOM	transportation command
UAS	unmanned aircraft system
UCT	underwater construction team
UBM	ultimate building machine
UFC®	Unified Facilities Criteria
UFS	universal fabric structures
UI	unit of issue
UJTL	Universal Joint Task List
U.S.	United States
USACE	United States Army Corps of Engineers
USACHPPM	United States Army Center for Health Promotion and Preventive Medicine
USAES	United States Army Engineer School
USAID	United States Agency for International Development

USAR	United States Army Reserve
USAES	United States Army Engineer School
USAREUR	United States Army, European Command
USC	United States Code
USCENTCOM	United States Central Command
USEUCOM	United States European Command
USNORTHCOM	United States Northern Command
USPACOM	United States Pacific Command
USSOUTHCOM	United States Southern Command
USTRANSCOM	United States Transportation Command
UTC	unit type code
UXO	unexploded explosive ordnance
vac	volts, alternating current
ver	vertical
w	water
w/	with
WAAS	Wide Area Augmentation System
WARNORD	warning order
WDRT	water detection response team
Wi-Fi®	wireless fidelity
WO1	Warrant Officer 1
WO2	Warrant Officer 2
WRDB	Water Resources Database
WRDT	water resource detection team
yd	yard

SECTION II – TERMS

*airfield damage repair

Encompasses all actions required to repair airfield and landing zone operating surfaces and infrastructure or services to conduct operations at a base or location seized from the enemy or offered for use by a host nation. It also includes repairs required to sustain operations or to reestablish operations after enemy attack at an airfield. Also called **ADR**.

area damage control

(joint/NATO) Measures taken before, during, or after hostile action or natural or manmade disasters, to reduce the probability of damage and minimize its effects. Also called **ADC**. (JP 3-10)

assured mobility

Actions that give the force commander the ability to maneuver where and when he desires without interruption or delay to achieve the mission. (FM 3-34)

assured mobility (fundamentals)

Predict, detect, prevent, avoid, neutralize and protect. These fundamentals support the implementation of the assured mobility framework. (FM 3-34)

bare base

A base having minimum essential facilities to house, sustain, and support operations to include, if required, a stabilized runway, taxiways, and aircraft parking areas. A bare base must have a source of water that can be made potable. Other requirements to operate under bare base conditions form a necessary part of the force package deployed to the bare base. See also **base.** (JP 3-05.1)

civil engineering

(joint) Those combat support and combat service support activities that identify, design, construct, lease, or provide facilities and which operate, maintain, and perform war damage repair and other engineering functions in support of military operations (JP 1-02)

civil support

Department of Defense support to U.S. civil authorities for domestic emergencies and for designated law enforcement and other activities. Also called **CS**. (JP 3-26) See FM 1.

engineer functions

The three engineer functions include combat, general, and geospatial engineering. (FM 3-34)

Engineer Regiment

All active component and reserve component engineer organizations (as well as the Department of Defense civilians and affiliated contractors and agencies within the civilian community) with a diverse range of capabilities that are all focused toward supporting the Army and its warfighting mission. (FM 3-34)

engineer work line

A coordinated boundary or phase line used to compartmentalize an area of operations (AO) to indicate where specific engineer units have primary responsibility for the engineer effort. It may be used at division level to discriminate between an AO supported by division engineer assets and an AO supported by direct support or general support corps engineer units. Also called **EWL**. (FM 5-100)

environmental baseline survey

A coordinated boundary or phase line used to compartmentalize an area of operations (AO) to indicate where specific engineer units have primary responsibility for the engineer effort. It may be used at division level to discriminate between an AO supported by division engineer assets and an AO supported by direct support or general support corps engineer units. Also called **EWL**. (FM 5-100)

general engineering

Those engineering capabilities and activities, other than combat engineering, that modify, maintain, or protect the physical environment. Examples include: the construction, repair, maintenance, and operation of infrastructure, facilities, lines of communication and bases; terrain modification and repair; and selected explosive hazard activities. Also called **GE**. (JP 3-34)

homeland defense

The protection of United States sovereignty, territory, domestic population, and critical defense infrastructure against external threats and aggression or other threats as directed by the President. Also called **HD**. (JP 3-27)

homeland security

A concerted national effort to prevent terrorist attacks within the United States; reduce America's vulnerability to terrorism, major disasters, and other emergencies; and minimize the damage and recover from attacks, major disasters, and other emergencies that occur. Also called **HS**. (JP 3-28)

operational environment

A composite of the conditions, circumstances, and influences that affect the employment of military forces and bear on decisions of the commander. (JP 3-0)

survivability

(joint) Concept which includes all aspects of protecting personnel, weapons, and supplies while simultaneously deceiving the enemy. Survivability tactics include building a good defense; employing frequent movement; using concealment, deception, and camouflage; and constructing fighting and protective positions for both individuals and equipment. [Note: The Army definition adds, "Encompasses planning and locating position sites, designing adequate overhead cover, analyzing terrain conditions and construction materials, selecting excavation methods, and countering the effects of direct and indirect fire weapons."] See FM 5-103. (Marine Corps) The degree to which a system is able to avoid or withstand a manmade hostile environment without suffering an abortive impairment of its ability to accomplish its designated mission. (FM 1-02)

survivability operations

The development and construction of protective positions, such as earth berms, dug-in positions, overhead protection, and countersurveillance means, to reduce the effectiveness of enemy weapon systems. (FM 5-103)

tele-engineering

Assists engineers and the commanders they support in planning and executing their missions with capabilities inherent in field force engineering (FFE) through exploitation of the Army's command, control, and communications architectures to provide a linkage between engineers and the appropriate nondeployed subject matter experts (SMEs) for resolution of technical challenges. Tele-engineering is under the proponency of the USACE.

time-phased force and deployment list

(joint) Appendix 1 to appendix A of the operation plan. It identifies types and/or actual units required to support the operation plan and indicates origin and ports of debarkation or ocean area. It may also be generated as a computer listing from the time-phased force and deployment data. Also called **TPFDL**. See FM 100-7. (JP 4-01.5)

Universal Joint Task List

A menu of capabilities (mission-derived tasks with associated conditions and standards, i.e., the tools) that may be selected by a joint force commander to accomplish the assigned mission. Once identified as essential to mission accomplishment, the tasks are reflected within the command joint mission essential task list. Also called **UJTL**. (JP 3-33)

This page intentionally left blank.

References

SOURCES USED

These are the sources quoted or paraphrased in this publication.

Air Force Publications

AFM 91-201, *Explosives Safety Standards.* 17 November 2008.

AFPAM 10-1403, *Air Mobility Planning Factors.* 18 December 2003.

ETL 04-7, *C-130 and C-17 Landing Zone (LZ) Dimensional, Marking, and Lighting Criteria.*
   29 March 2004.

ETL 97-9, *Criteria and Guidance for C-17 Contingency and Training Operations on Semi-Prepared
   Airfields.* 25 November 1997.

Army Publications

AR 40-5, *Preventive Medicine.* 25 May 2007.

AR 190-11, *Physical Security of Arms, Ammunition and Explosives.* 15 November 2006.

AR 405-10, *Acquisition of Real Property and Interests Therein.* 15 July 1974.

AR 415-16, *Army Facilities Components System.* 17 March 1989.

AR 420-1, *Army Facilities Management.* 12 February 2008.

AR 715-9, *Contractors Accompanying the Force.* 29 October 1999.

DA Pamphlet 40-11, *Preventive Medicine.* 22 July 2005.

DA Pamphlet 190-12, *Military Working Dog Program.* 30 September 1993.

DA Pamphlet 350-38, *Standards in Training Commission.* 24 July 2008.

DA Pamphlet 385-64, *Ammunition and Explosives Safety Standards.* 15 December 1999.

FM 1, *The Army.* 14 June 2005.

FM 3-0, *Operations.* 27 February 2008.

FM 3-05.40, *Civil Affairs Operations.* 29 September 2006.

FM 3-07, *Stability Operations.* 6 October 2008.

FM 3-13, *Information Operations: Doctrine, Tactics, Techniques, and Procedures.*
   28 November 2003.

FM 3-19.40, *Internment/Resettlement Operations.* 4 September 2007.

FM 3-34, *Engineer Operations.* 2 January 2004.

FM 3-34.2, *Combined-Arms Breaching Operations.* 31 August 2000.

FM 3-34.170, *Engineer Reconnaissance.* 25 March 2008.

FM 3-34.343, *Military Nonstandard Fixed Bridging.* 12 February 2002.

FM 3-34.465, *Quarry Operations.* 15 April 2005.

FM 3-34.468, *Seabee Quarry Blasting Operations and Safety Manual.* 19 December 2003.

FM 3-34.480, *Engineer Prime Power Operations.* 4 April 2007.

FM 3-90, *Tactics.* 4 July 2001.

FM 3-100.21, *Contractors on the Battlefield.* 3 January 2003.

FM 4-01.41, *Army Rail Operations.* 12 December 2003.

FM 4-30.13, *Ammunition Handbook: Tactics, Techniques, and Procedures for Munitions Handlers.*
   1 March 2001.

FM 5-0, *Army Planning and Orders Production.* 20 January 2005.

FM 5-34, *Engineer Field Data.* 19 July 2005.

FM 5-71-100, *Division Engineer Combat Operations.* 22 April 1993.

FM 5-100-15, *Corps Engineer Operations.* 6 June 1995.

FM 5-103, *Survivability.* 10 June 1985.

FM 5-116, *Engineer Operations: Echelons Above Corps (S&I, CDR, USATSC, ATTN: ATIC-TMSD-T).* 9 February 1999.

FM 5-134, *Pile Construction.* 18 April 1985.

FM 5-277, *M2 Bailey Bridge.* 9 May 1986.

FM 5-410, *Military Soils Engineering.* 23 December 1992.

FM 5-412, *Project Management.* 13 June 1994.

FM 5-415, *Fire-fighting Operations.* 9 February 1999.

FM 5-424, *Theater of Operations Electrical Systems.* 25 June 1997.

FM 5-428, *Concrete and Masonry.* 18 June 1998.

FM 5-436, *Paving and Surfacing Operations.* 28 April 2000.

FM 5-480, *Port Construction and Repair.* 12 December 1990.

FM 6-0, *Mission Command: Command and Control of Army Forces.* 11 August 2003.

FM 7-15, *The Army Universal Task List.* 31 August 2003.

FM 7-100, *Opposing Force Doctrinal Framework and Strategy.* 1 May 2003.

FM 10-52, *Water Supply in Theaters of Operations.* 11 July 1990.

FM 10-52-1, *Water Supply Point Equipment and Operations.* 18 June 1991.

FM 10-67, *Petroleum Supply in Theaters of Operations.* 18 February 1983.

FM 10-67-1, *Concepts and Equipment of Petroleum Operations.* 2 April 1998.

FM 20-3, *Camouflage, Concealment, and Decoys.* 30 August 1999.

FM 27-10, *The Law of Land Warfare.* 18 July 1956.

FM 34-130, *Intelligence Preparation of the Battlefield.* 8 July 1994.

FM 4-20.07, *Quartermaster Force Provider Company.* 29 August 2008.

FM 55-60, *Army Terminal Operations.* 15 April 1996.

FM 100-10-1, *Theater Distribution.* 1 October 1999.

FM 100-16, *Army Operational Support.* 31 May 1995.

TM 5-235, *Special Surveys.* 18 September 1964.

TM 5-300, *Real Estate Operations in Oversea Commands.* 10 December 1958.

TM 5-301-1, *Army Facilities Components System--Planning (Temperate).* 27 June 1986.

TM 5-301-2, *Army Facilities Components System--Planning (Tropical).* 27 June 1986.

TM 5-301-3, *Army Facilities Components System--Planning (Frigid).* 27 June 1986.

TM 5-301-4, *Army Facilities Components System--Planning (Desert).* 27 June 1986.

TM 5-302-1, *Army Facilities Components System: Design (S&I, USAEDH, Attn: HNDED-FD, Huntsville, AL 35807-4301).* 28 September 1973.

TM 5-302-2, *Army Facilities Components System: Design (S&I, USAEDH, Attn: HNDED-FD, Huntsville, AL 35807-4301).* 28 September 1973.

TM 5-302-3, *Army Facilities Components System: Design (S&I, USAEDH, Attn: HNDED-FD, Huntsville, AL 35807-4301).* 28 September 1973.

TM 5-302-4, *Army Facilities Components System: Design (S&I, USAEDH, Attn: HNDED-FD, Huntsville, AL 35807-4301).* 28 September 1973.

TM 5-302-5, *Army Facilities Components System: Design* (S&I, USAEDH, Attn: HNDED-FD, Huntsville, AL 35807-4301). 28 September 1973.

TM 5-303, *Army Facilities Components System - Logistic Data and Bills of Materiel.* 1 June 1986.

TM 5-315, *Firefighting and Rescue Procedures in Theaters of Operations.* 20 April 1971.

TM 5-5420-212-10-1, *Operator's Manual for Medium Girder Bridge Including Bridge Set (NSN 5420-00-172-3520) Bridge Erection Set (5420-00-172-3519) Link Reinforcement Set (5420-01-139-1503).* 16 February 1993.

TM 5-5420-278-10, *Operator's Manual for Improved Ribbon Bridge (IRB), Ramp Bay M16 (NSN 5420-01-470-5825), P/N 12478918 (EIC: XMT), Interior Bay M17 (NSN 5420-01-470-5824) P/N 12478919 (EIC: XMS).* 8 April 2003.

TM 5-5420-279-10, *Operator Manual for Dry Support Bridge (DSB) (NSN 5420-01-469-7479).* 10 May 2004.

TM 5-820-1, *Surface Drainage Facilities for Airfields and Heliports.* 20 August 1987.

Joint Publications

JP 1-02, *Department of Defense Dictionary of Military and Associated Terms.* 12 April 2001.

JP 3-0, *Joint Operations.* 17 September 2006.

JP 3-26, *Homeland Security.* 2 August 2005.

JP 3-34, *Joint Engineer Operations.* 12 February 2007.

JP 4-03, *Joint Bulk Petroleum and Water Doctrine.* 23 May 2003.

JP 5-0, *Joint Operation Planning.* 26 December 2006.

Miscellaneous

AFH 32-1084, *Civil Engineering,* Category Code 141-182, *Hardened Aircraft Shelters.* 1 September 1996.

Base Camp Facilities Handbook, September 2005.

CJCSM 3122.03C, *Joint Operation Planning and Execution System Volume II: Planning Formats and Guidance.* 17 August 2007.

CJCSM 3500.04E, *Universal Joint Task Manual.* 25 August 2008.

DOD Directive 4165.6, *Real Property.* 13 October 2004.

DOD Directive 4525.6-M, *Department of Defense Postal Manual.* 15 August 2002.

DOD Instruction 4715.5, *Management of Environmental Compliance at Overseas Installations.* 22 April 1996.

DOD Instruction 4715.05-G, *Overseas Environmental Baseline Guidance Document.* 1 May 2007.

DOD Instruction 6055.6, *DOD Fire and Emergency Services (F&ES) Program.* 21 December 2006.

Environmental Baseline Survey Handbook: *Contingency Operations (Overseas),* September 2005.

EO 12114, *Environmental Effects Abroad of Major Federal Actions.* 4 January 1979.

EO 13148, *Greening the Government through Leadership in Environmental Management.* 21 April 2000.

Geneva Convention, 1949.

Geneva Convention IV, *Relative to the Protection of Civilian Persons in Time of War.* 12 August 1949.

Hague Convention, 18 October 1907.

Hague Convention, *Relative to the Protection of Cultural Property in the Event of Armed Conflict.* 14 May 1954.

JP 4-04, *Joint Doctrine for Civil Engineering Support.* 26 September 1995.

National Electrical Safety Code® 2007 Handbook, 2 Edition, *McGraw-Hill Professional.* 19 October 2006.

National Electrical Code® 2008 Handbook, 1 Edition, *CENGAGE Delmar Learnin.* 10 January 2008.

National Environmental Policy Act, 1 January 1970.

ST 20-23-8, *Use of Demining Dogs and Military Operations,* September 2003.

TB MED 577, *Sanitary Control and Surveillance of Field Water Supplies.* 15 December 2005.

TG 248, *Guide for Deployed Preventive Medicine Personnel on Health Risk Management.* August 2001.

Title 10, USC Armed Forces, Section 401, *Humanitarian and Civic Assistance Provided in Conjunction with Military Operations.*

UFC 3-240-10A, *Sanitary Landfill.* 16 January 2004.

UFC 3-260-01, *Airfield and Heliport Planning and Design.* 17 November 2008.

UFC 3-260-02, *Pavement Design for Airfields.* 30 June 2001.

UFC 3-260-03, *Airfield Pavement Evaluation.* 15 April 2001.

UFC 3-270-07, *O&M: Airfield Damage Repair.* 12 August 2002.

UFC 4-141-10N, *Design: Aviation Operation and Support Facilities.* 16 January 2004.

UFC 4-150-02, *Dockside Utilities for Ship Service.* 12 May 2003.

UFC 4-150-06, *Military Harbors and Coastal Facilities.* 12 December 2001.

UFC 4-150-07, *Maintenance and Operation: Maintenance of Waterfront Facilities.* 19 June 2001.

UFC 4-150-08, *Inspection of Mooring Hardware.* 1 April 2001.

UFC 4-151-10, *General Criteria for Waterfront Construction.* 10 September 2001.

USAREUR, *Base Camp Facility Standards for Contingency Operations.* 1 February 2004.

USCENTCOM Sand Book, *Construction and Base Camp Development in the USCENTCOM Area of Responsibility.* 1 December 2004.

Multi-Service Publications

FM 1-02/MCRP 5-12A, *Operational Terms and Graphics.* 21 September 2004.

FM 3-90.12/MCWP 3-17.1, *Combined Arms Gap-Crossing Operations.* 1 July 2008.

FM 3-100.4/MCRP 4-11B, *Environmental Considerations in Military Operations.* 15 June 2000.

FM 5-34/MCRP 3-17A, *Engineer Field Data.* 19 July 2005.

FM 5-430-00-1/AFPAM 32-8013, Volume I, *Planning and Design of Roads, Airfields, and Heliports in the Theater of Operations – Road Design.* 26 August 1994.

FM 5-430-00-2/AFJPAM 32-8013, Volume II, *Planning and Design of Roads, Airfields, and Heliports in the Theater of Operations – Airfield and Heliport Design.* 29 September 1994.

FM 5-472/NAVFAC MO 330/AFJMAN 32-1221(I), *Materials Testing.* 27 October 1999.

FM 5-484/NAVFAC P-1065/AFMAN 32-1072, *Multiservice Procedures for Well-Drilling Operations.* 8 March 1994.

NAVFAC MO-213/AFR 91-8/TM 5-634, *Solid Waste Management.* May 1990.

NWP 4-04.1/MCWP 4-11.5, *Seabee Operations in the MAGTF.* November 1997.

TM 5-610, *Preventive Maintenance for Facilities Engineering, Buildings and Structures.* 1 November 1979.

TM 5-683/NAVFAC MO-116/AFJMAN 32-1083, *Facilities Engineering: Electrical Interior Facilities.* 15 December 1995.

TM 5-684/NAVFAC MO-200/AFJMAN 32-1082, *Facilities Engineering - Electrical Exterior Facilities.* 29 November 1996.

TM 5-820-1/AFM 88-5, *Surface Drainage Facilities for Airfields and Heliports.* 20 August 1987.

Navy Publications

NWP 4-11, *Environmental Protection.* March 1999.

OPNAVINST 5090.1C, *Environmental Readiness Program Manual.* 30 October 2007.

OPNAVINST 11300.5C, *Mobile Utilities Support Equipment (MUSE) Program.* 14 May 2007.

Standardization Agreement

STANAG 2010 (Ed 5), *Military Load Classification Markings.* 18 November 1980.

STANAG 2021(Ed 5), *Military Computation of Bridge, Ferry, Raft, and Vehicle Classifications.* 18 September 1990.

Quadripartite Standardization Agreement

QSTAG-180 (Ed 4), *Military Load Classification of Bridges (Computation of Bridge, Ferry, Raft, and Vehicle Classifications.* 11 August 1986.

DOCUMENTS NEEDED

These documents must be available to the intended users of the publication.

DA Form 2028, *Recommended Changes to Publications and Blank Forms.*

DD Form 1354, *Transfer and Acceptance of Military Real Property.*

DD Form 1391, *FY, Military Construction Project Data.*

DA Forms are available on the APD web site (www.apd.army.mil). DD forms are available on the OSD web site (www.dtc.mil/whs/directives/infomgt/forms/formsprogram.htm).

READINGS RECOMMENDED

These sources contain relevant supplemental information.

None

This page intentionally left blank.

Index